SpringerBriefs in Physics

SpringerBriefs in Physics are a series of slim high-quality publications encompassing the entire spectrum of physics. Manuscripts for SpringerBriefs in Physics will be evaluated by Springer and by members of the Editorial Board. Proposals and other communication should be sent to your Publishing Editors at Springer.

Featuring compact volumes of 50 to 125 pages (approximately 20,000–45,000 words), Briefs are shorter than a conventional book but longer than a journal article. Thus, Briefs serve as timely, concise tools for students, researchers, and professionals.

Typical texts for publication might include:

- A snapshot review of the current state of a hot or emerging field
- A concise introduction to core concepts that students must understand in order to make independent contributions
- An extended research report giving more details and discussion than is possible in a conventional journal article
- A manual describing underlying principles and best practices for an experimental technique
- An essay exploring new ideas within physics, related philosophical issues, or broader topics such as science and society

Briefs allow authors to present their ideas and readers to absorb them with minimal time investment. Briefs will be published as part of Springer's eBook collection, with millions of users worldwide. In addition, they will be available, just like other books, for individual print and electronic purchase. Briefs are characterized by fast, global electronic dissemination, straightforward publishing agreements, easy-to-use manuscript preparation and formatting guidelines, and expedited production schedules. We aim for publication 8–12 weeks after acceptance.

Paul van der Schoot

Molecular Theory of Nematic (and Other) Liquid Crystals

An Introduction

with figures by Mariana Oshima Menegon

 Springer

Paul van der Schoot
Department of Applied Physics
Eindhoven University of Technology
Eindhoven, Noord-Brabant
The Netherlands

ISSN 2191-5423   ISSN 2191-5431   (electronic)
SpringerBriefs in Physics
ISBN 978-3-030-99861-5   ISBN 978-3-030-99862-2   (eBook)
https://doi.org/10.1007/978-3-030-99862-2

This Springer imprint is published by the registered company Springer Nature Switzerland AG
The registered company address is: Gewerbestrasse 11, 6330 Cham, Switzerland

Preface

Liquid crystals are complex fluids that exhibit some sort of long-range molecular order and internal structure. This internal molecular structure is not static but pliable, and responds to external perturbation such as that caused by an applied electric field or by the presence of the walls of the container holding the material. Liquid crystals actually have a sense of direction and are, therefore, *not* isotropic and hence *not* like ordinary fluids. Consequently, the physical properties of liquid crystals are *not* the same in all directions, that is, *anisotropic*. It is on account of their unusual anisotropic dielectric and optical properties, and their responsiveness to external fields, that liquid crystals have found a myriad of applications in opto-electronic technologies, including displays, optical switches and switchable ("smart") glass.

In the light of this, we should not be surprised that there is a considerable techno-logical interest in liquid crystals. As a matter of fact, these materials have attracted immense interest in the scientific community too, exactly because they have such unusual properties [1]. This, in turn, has led the topic of liquid crystals to enter, in one form or another, in university curricula in chemistry, chemical engineering, materials science, applied and theoretical physics, and even mathematics. Typically, university courses that cover aspects of the chemistry, physics and applications of liquid crystals are at the graduate (master) student level.

There are very many excellent graduate-level textbooks specialising in liquid crystal science. Some offer more advanced-level material than others, but what most of them share is an aim to be comprehensive and as a result cover a large number of topics. Focusing on the materials science spectrum of these texts, this usually entails a discussion of many aspects of the physics of more than a few of the many kinds of liquid crystal that have been identified since their discovery in the late nineteenth century [2]. This includes discussions of the thermodynamics, elasticity, flow behaviour, optical properties, device applications and so on. To a lesser extent, this is also true for most review papers [3, 4], including the more didactic ones [5].

Most texts present, in some form, the basic molecular theories describing the transition between the "conventional" isotropic fluid and, in a way, the simplest and arguably most widely investigated liquid crystalline state, known as the *nematic liquid crystal*. The molecules or particles that make up the nematic liquid-crystalline fluid are aligned along some preferred direction and yet exhibit no long-range positional order such as that found in crystals. The spontaneous alignment of the particles in the fluid gives rise to the highly complex elastic and flow behaviour so characteristic of liquid crystals in general.

What causes the molecules or particles to spontaneously align depends on the type of liquid crystal at hand. We can distinguish between two kinds of liquid crystal, and consequently also between two kinds of *nematic* liquid crystal. These are commonly referred to as *thermotropic* and *lyotropic* nematics. Thermotropic liquid crystals are fluid phases of low molecular weight compounds, whereas lyotropic liquid crystals are dispersions of colloidal particles in a host fluid, usually implied to be isotropic. Colloidal particles are often solid particles of which the dimensions range from a few nanometres to a few micrometres.[1] The thermodynamic driving force for spontaneous molecular alignment in thermotropic nematics is a trade-off between enthalpy and entropy, and that in lyotropic nematics a trade-off between different forms of entropy. This make the former respond to changes in temperatures, whilst for the latter it is the particle concentration that, by and large, determines what kind of stable phase presents itself.

Theories describing the isotropic-to-nematic phase transition can be formulated for both types of system, and, as a matter of fact, for other liquid crystal transitions too. The two prototypical *molecular* theories are *Maier-Saupe theory* for thermotropic nematics and *Onsager theory* for lyotropic nematics. These two theories are usually treated next to each other in textbooks, without any discussion or explanation of the underlying connection between them. This might have to do with the circumstance that the thermotropic and lyotropic liquid crystal communities have been more or less disconnected for a century or so. As these communities have recently started to integrate [6], it would seem opportune to offer a graduate-level text that highlights the similarities rather than the differences between these theories.

As shall become clear in this introductory text, both theories are rooted in (and can be derived from) the same formalism known as (classical) *density functional theory*. It may seem unusual to take what is often seen as a rather advanced theoretical framework in an introductory text, in order to explain the underlying physics of the spontaneous ordering of particles. However, density functional theory is actually rather intuitive, and one of the aims of this text is indeed to demystify it and make it palatable to a wider audience. In my view, the theory need not be rigorous to provide physical insight. This insight hopefully lowers the barrier to reading more advanced works in liquid-state and soft-matter theory. For this reason, all chapters end with suggestions to additional, more advanced reading material.

Most of the material presented in this *SpringerBrief* was taught internationally to graduate student audiences of quite diverse background, audiences that included chemists, chemical engineers and physicists. Hence, the target audience of the book

---

[1] Sometimes the particles are self-assembled from smaller molecular building blocks, such as surfactants. In that case, the particles themselves, which are known as micelles, are fluid-like. Liquid crystals of self-assembled particles are sometimes exclusively referred to as lyotropic liquid crystals.

is expressly *not* restricted to theoreticians or to graduate students, and specifically includes the entire experimental soft matter community. The suggested reading material ranges from introductory to quite advanced, to fit as wide an audience as possible.

The only prior background knowledge required is elementary calculus and a basic notion of thermodynamics and probability theory. Most mathematical tricks not dealt with in detail in the main text are covered in the problems at the end of each chapter. Solutions to the problems can be found at the end of this book. Having taken an introductory statistical mechanics or statistical thermodynamics course is not necessary albeit that it would obviously not hurt. The same is true for any prior background knowledge in soft matter physics or physical chemistry or colloid science.

Readers wishing to refresh their memory of the basic principles of statistical mechanics may find the introductory textbooks of Ben Widom [7] and David Chandler [8] very pedagogical. The latter also revisits the basics of thermodynamics in a compact and enlightening way. For a pedagogical primer in soft matter physics and density functional theory, the reader is referred to the textbook by Jean-Luis Barrat and Jean-Pierre Hansen [9] and that by Masao Doi [10]. More advanced discussions of liquid-state theory and density functional theory can be found in the book by Jean-Pierre Hansen and I. R. McDonald [11].

The mathematically insecure may find Michael Stone and Paul Goldbart's book on mathematics for physics [12] useful, or consult the mathematics for scientists and engineers tome of Donald McQuarrie [13]. For those interested in the history of liquid crystal science and technology are referred to the wonderful book of David Dunmur and Tim Sluckin [2]. Finally, the coming together of the fields of colloids and thermotropic liquid crystals leads to beautiful and exciting new physics may be seen in the book edited by Jan Lagerwall and Chiusy Scalia [6].

It is important to stress that the literature reference list provided at the end of this book is not meant to be complete and, in fact, is not at all complete. The list is merely intended to provide a helping hand in finding relevant literature, a stepping stone to the very extensive body of literature dedicated to the field. The various review papers cited in this book provide an extensive introduction to the world of thermotropic and lyotropic liquid crystals.

Finally, I thank Joost de Graaf, Thijs van der Heijden, Mariana Oshima Menegon and Shari Patricia Finner for a critical reading of (early versions of) the manuscript, corrections and suggested improvements. I am grateful to Prof. Monika Marzec for her permission to reproduce the polarisation micrograph of 5CB, and to Profs. Seth Fraden (Brandeis University, Waltham, USA) and Zvonimir Dogic (University of California Santa Barbara, USA) for the photographic images of the co-existing isotropic and nematic phases of TMV. This project has received funding from the European Union Horizon 2020 research and innovation programme under the Marie Skłodowska-Curie grant agreement No. 641839.

Eindhoven, The Netherlands   Paul van der Schoot
January 2022

References

1. P. G. de Gennes and J. Prost, *The Physics of Liquid Crystals*, Second Edition (OUP, Oxford, 1993).
2. D. Dunmur and T. Sluckin, *Soap, and flat-screen TVs* (OUP, Oxford, 2011).
3. M. J. Stephen and J. P. Straley, *Physics of liquid crystals*, Rev. Mod. Phys. **46** (1974), 617.
4. G. J. Vroege and H. N. W. Lekkerkerker, *Phase transitions in lyotropic colloidal and polymer liquid crystals*, Rep. Prog. Phys. **55** (1992), 1241.
5. D. Andrienko, *Introduction to liquid crystals*, J. Mol. Liq. **267** (2018), 520.
6. J. P. F. Lagerwall and G. Scalia, eds., *Liquid Crystals with Nano and Microparticles*, vol. I and II, (World Scientific, New Jersey, 2017).
7. B. Widom, *Statistical mechanics - a concise introduction for chemists* (CUP, Cambridge, 2002);
8. D. Chandler, *Introduction to modern statistical mechanics* (OUP, Oxford, 1987).
9. J.-L. Barrat and J.-P. Hansen, *Basic concepts for simple and complex fluids* (CUP, Cambridge, 2003).
10. M. Doi, *Soft matter physics* (OUP, Oxford, 2013).
11. J.-P. Hansen and I.R. McDonald, *Theory of simple liquids - with applications to soft matter*, 4th edition (AP, Amsterdam, 2013).
12. M. Stone and P. Goldbart, *Mathematics for physics* (CUP, Cambridge, 2000).
13. D. A. McQuarrie, *Mathematical methods for scientists and engineers* (USB, Sausalito, 2003).

Acknowledgement

Most images in this book were kindly created by Dr. Mariana Oshima Menegon.

Contents

Chapter 1
Introduction

Abstract This chapter reviews the basic differences between gases, liquids, and solids, the transitions between them and the importance of symmetry. Some attention is given to the earliest equation of state put forward for non-ideal gases that is based on a molecular view of matter: the van der Waals equation of state. This equation of state predicts the condensation of a gas into a liquid and the existence of a critical point. The concept of universality emerges naturally from it.

Very few people are not familiar with the three main aggregated states of matter: the gas, the liquid and the crystalline solid. A gas has a low density and does not keep its volume or shape: it expands to fill the container it is confined to. Liquids and solids are both dense phases, where the former keeps its volume but not its shape whilst the latter keeps both in any shape or size of the container. Gases and liquids flow if a force is exerted onto them and exhibit viscous flow behaviour, implying they can be poured. Indeed, a relatively dense gas can be contained, for a while at least, in a beaker and poured out of that beaker. Solids deform under mechanical loading but do not flow and, in principle, only exhibit elastic deformation behaviour. Upon removal of the external force, they relax back to their initial shape.

Solids are crystals and are characterised by so-called *long-range positional order*. This is because the atoms and molecules that make up crystals are not roaming freely in space, but are associated with essentially fixed positions of an underlying crystal lattice. Gases and liquids do not exhibit any kind of long-range positional order. This means that the particles can move about, and eventually visit all positions within the confines of their allowed space, that is, their allowed volume, with equal probability. The main difference between a gas and a liquid is the density, which can differ by a factor of as much as one thousand even under conditions of co-existence. The main difference between a crystal, on the one hand, and a gas or liquid, on the other, is the degree of order, or, in other words, their *symmetry*. See Fig. 1.1.

Loosely speaking, symmetry refers to the number of ways an object can be superimposed on itself by translation, rotation, or reflection without changing the way that it looks. Because of their regular structure, crystals have a lower symmetry than gases and liquids, which are *isotropic*. This means that their physical properties,

P. van der Schoot, *Molecular Theory of Nematic (and Other) Liquid Crystals*,
SpringerBriefs in Physics, https://doi.org/10.1007/978-3-030-99862-2_1

Fig. 1.1 Two dimensional representation of a gas (left), liquid (middle) and solid phase (right).

such as their dielectric and mechanical response, do *not* depend on the direction in which we probe them. Crystals are not like that at all: their properties depend on the direction of investigation. Crystals are, by virtue of their crystal symmetry, of which there are very many different kinds, *anisotropic*. Because crystals have a lower symmetry than the fluids they emerge from, the freezing transition is an example of what is called a *symmetry-breaking transition*.

Figure 1.2 shows a generic phase diagram of atoms, small molecules and spherical colloidal particles that sometimes are seen as large model atoms, represented in the left figure in the pressure-temperature or $p - T$ plane, and in the right figure in the temperature-density or $T - \rho$ plane, where the density $\rho = N/V$ refers to the number of particles N in the volume V of the material. The drawn lines in the pressure-temperature diagram (Fig. 1.2, left) indicate *discontinuous* (or *first order*) transitions between the different states of matter. They are called discontinuous transitions because the appropriate physical quantity describing the state of the material jumps in a discontinuous manner from one value to another when going from one phase to the other. This can be the density or the degree of crystalline order.

These lines represent a continuum of conditions where phases co-exist and are in stable thermodynamic equilibrium with each other. They are known as the *sublimation line* for the gas-solid co-existence, the *freezing* or *melting line* for the fluid-solid transition and the *boiling line* separating the liquid and gaseous phases. Under conditions of co-existence, phases have equal temperatures, chemical potentials and pressures. This is a consequence of the second law of thermodynamics, stating that an isolated system in thermodynamic equilibrium *maximises* its entropy [1].

Indicated in the phase diagrams of Fig. 1.2 are also the *triple point* (TP) at which a solid, a liquid *and* a gas co-exist, and the *critical point* (CP) at the end of the liquid-gas transition line. It is also sometimes referred to as the critical endpoint of the first order gas-liquid transition.[1] At fixed pressure p but increasing temperature T, the typical sequence of the aggregated state of matter goes from a crystalline phase, via a liquid phase, to a gas phase. This happens at temperatures above the triple point

---

[1] The modern classification of phase transitions revolves, loosely speaking, around whether or not two phases can co-exist or exhibit a latent heat for first order transitions, and divergent thermodynamic susceptibilities and an absence of a latent heat for second order ones.

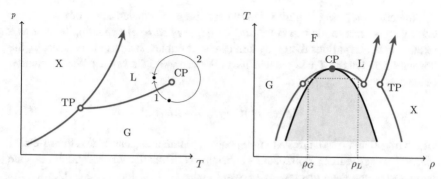

Fig. 1.2 Left: phase diagram of atoms, small molecules and colloidal particles in the pressure (p)—temperature (T) plane. Right: phase diagram in the temperature (T)—density (ρ) plane. Red lines are *binodals*, the black line is the liquid-gas *spinodal*. The horizontal, (dotted) *tie line* connects densities in co-existing gaseous and liquid phases with densities ρ_G and ρ_L. See also the main text. X: crystal, L: liquid, G: gas, F: super-critical fluid. Indicated are also the triple point, TP, and the critical point, CP. Gaseous and liquid states can be inter-converted via a discontinuous phase transition that crosses the boiling line (path 1), or by circumnavigating the critical point (path 2) in which case the conversion is smooth.

and below the critical point of a compound. There is *no* equivalent critical point for the melting transition, because it is forbidden by symmetry. This is represented by the arrow head connected to the crystal-fluid melting line in the phase diagram of Fig. 1.2.

Figure 1.2 (right) shows that the densities ρ_G and ρ_L of the co-existing gas and liquid-phases approach each other upon approaching the critical temperature from below. In the region above the critical point (CP) there is no liquid-gas transition and there is no difference between the liquid and the gaseous states of a compound, in which case it becomes *super critical*. The thermodynamic state of the material is then referred to as a *super-critical fluid*, or sometimes just *fluid* for short. Consequently, we can transform a liquid into a gas without a sudden phase change by simply circumnavigating the critical point, as is indicated in the Fig. 1.2 (left) by the paths 1 and 2. At the critical density the transition between fluid phase represents a continuous or second order transition, for the density difference between the coexisting gaseous and liquid phases can be made arbitrarily small.

Because temperatures must be equal under conditions of co-existence, we can draw *tie lines* in Fig. 1.2 (right) between the densities in the co-existing phases in the *phase gap*, separating the equilibrium densities in these phases. The collection of complementary pairs of gas and liquid density in the co-existing phases is called the *binodal*. There are also binodals separating the crystal phase and the co-existing gas, and the crystal phase and the liquid and fluid phases. The binodal of the fluid-crystal co-existence does not end in a critical point, which, as already alluded to, is forbidden for reasons of symmetry: a material cannot be isotropic and anisotropic at the same time.

Arguably, the first *molecular theory* describing the condensation of a gas is the one that gave rise to the *van der Waals equation of state*. An equation of state is a relation between various thermodynamic state variables, such as the pressure p, the absolute temperature T and the density ρ. We can write the van der Waals equation of state as

$$(p + a\rho^2)(1 - b\rho) = \rho k_B T, \tag{1.1}$$

where $a \geq 0$ and $b \geq 0$ are system-dependent parameters, and $k_B T$ is the so-called *thermal energy* with k_B Boltzmann's constant.[2] The van der Waals equation of state is, in a way, the ideal gas law, $p = \rho k_B T$, corrected for interactions between the particles that make up the gas. The parameter a accounts for attractive interactions between them, and b for repulsive interactions. The latter corrects for the fact that the total volume of a gas (or liquid) is not accessible to the centres of mass of the particles as they take up a finite volume. This explains why the correction is linear in the density ρ: it reflects the loss of so-called free volume, and hence of translational entropy, of the particles. Tacit assumption in Eq. (1.1) is the condition that $\rho b < 1$: the fluid state becomes crystalline for densities below that. See also exercise 3 of this chapter.

Attractive interactions cause particles to pair up and effectively reduce the number of independently moving particles in the gas. The probability that two particles pair up must be proportional to the square of the density: to find one particle in some small volume is proportional to ρ, and to find two in the same small volume must hence be proportional to ρ^2, at least if they are uncorrelated and behave independently of each other. One way to deal with this pairing up of particles is to reduce the pressure by an amount that is proportional to ρ^2. This is obviously an approximation because in reality particles *are* statistically correlated in space, even if this correlation does not go beyond a few particle diameters. Indeed, if they attract each other, the probability of finding a particle near another particle is greater than what you would expect based on the average density. The magnitude of the parameter a effectively accounts for this.

The van der Waals equation of state describes very different behaviour for high and low temperatures. Indeed, for temperatures T above a critical temperature T_c the pressure p is a single-valued function of the particle density ρ. See Fig. 1.3. If the temperature is exactly equal to the critical temperature, $T = T_c$, this curve remains a non-decreasing function of ρ, but has an inflection point, marked with a filled circle in the figure. Below the critical temperature, $T < T_c$, the pressure becomes multi-valued over some range of densities. It is in this range of densities (or pressures) that the liquid-gas transition takes place.

That this must be so, can in fact be concluded from what we know from the laws of thermodynamics. Indeed, it is a consequence of the second law of

---

[2] For systems described by classical (Boltzmann) statistics, the relevant energy scale is the thermal energy.

Fig. 1.3 Schematic of isotherms according to the van der Waals equation of state in the pressure (p) *versus* the density (ρ) plane. Indicated are isotherms above the critical temperature $T > T_c$, at the critical temperature $T = T_c$, and below the critical temperature $T < T_c$. The circle indicates the critical point. The dashed portion of the sub-critical isotherm indicates where the fluid is thermodynamically unstable.

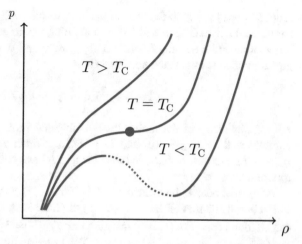

thermodynamics that homogeneous states with a negative compressibility cannot be (thermodynamically) stable. Consequently, we have to insist that $\partial p/\partial \rho > 0$. This immediately allows us to pinpoint the conditions of *marginal thermodynamic stability*, collectively known as the *spinodal*, and set by the condition $\partial p/\partial \rho = 0$. It is indicated in Fig. 1.1 (right).

The critical temperature is the temperature above which the spinodal ends: this is the highest temperature for which the pressure has an inflection point as a function of the density: $\partial p/\partial \rho = \partial^2 p/\partial \rho^2 = 0$. For the van der Waals equation of state, Eq. (1.1), we obtain $k_B T_c = a(2/3)^3/b$, $\rho_c = 1/3b$ and $p_c = a/27b^2$ in terms of the parameters a and b. According to the van der Waals equation of state, we have exactly at the critical point the relation $p_c/\rho_c k_B T_c = 3/8 = 0.325$, irrespective of the chemical compound! This agrees reasonably well with experimental values that are typically are in the range 0.21–0.31, with an average value of 0.29 [2].

To calculate the gas-liquid binodals, in other words, locate phase equilibria between gaseous and liquid states, we need to equate the temperatures T, pressures p and chemical potentials μ of the two states, as already alluded to. The van der Waals equation of state provides sufficient information to be able to do this. To equate chemical potentials, we need to make use of what is known as the *Maxwell construction* or the *equal area construction*, alternatively, of the *common tangent construction*. The latter method we discuss in exercise 3 below. The former relies on making use of a so-called Maxwell relation between the pressure and chemical potential of a compound. We shall not dwell on it here.

The great triumph of the van der Waals equation of state is not just that it predicts a phase transition between a gaseous and a liquid state of matter, but that it can explain the experimentally observed *law of corresponding states*. The law of corresponding states puts forward that the liquid-gas binodal is *universal* if expressed in terms of appropriately *reduced units*. "Universal" here means that it is valid for in principle *any* kind of atom or small molecule, and "appropriately reduced units" refers to the thermodynamic variables, T, ρ, and p being scaled to (or divided by) their values at the critical point, which we denote T_c, ρ_c, and p_c. In terms of the reduced units $T_r \equiv T/T_c$, $\rho_r \equiv \rho/\rho_r$ and $p_r \equiv p/p_c$, the van der Waals equation of state attains the universal form

$$(p_r + 3\rho_r^2)(3 - \rho_r) = 8\rho_r T_r. \tag{1.2}$$

All the chemistry of the molecules is embedded in the phenomenological model parameters $a = 9k_B T_c/8\rho_c$ and $b = 1/3\rho_c$, which are measurable because p_c, ρ_c, and Tc are observables. Values of a and b have been tabulated for very many different substances.[3]

At, or just below, the critical point, so for $T_r \uparrow 1$,[4] the difference of the density of the liquid and that of the co-existing gas phase is small. In that case, the binodal can be found from Eq. (1.2) by insisting that $T_r^{liquid} = T_r^{gas} = T_r$, $p_r^{liquid} = p_r^{gas}$ and that $\rho_r^{gas} = 1 - \varepsilon$ and $\rho_r^{liquid} = 1 + \varepsilon$. The latter set of equations replaces equality of chemical potentials. We solve the governing equation for ε by making use of a Taylor expansion for $|\varepsilon| \ll 1$. (See also exercise 1 of this chapter.)

After a little algebra, we find the binodal to obey (in reduced units) the following limiting expression[5]

$$\left(\rho_r^{liquid} - \rho_r^{gas}\right) \sim 4\sqrt{1 - T_r}, \tag{1.3}$$

according to the van der Waals equation of state. Hence, the phase gap widens with the one-half power of the distance in temperature to the critical temperature. The power-law exponent is called the *critical exponent*, and the value of one-half is somewhat larger than the experimental value of about one-third sufficiently close to the critical point [2]. This means that the van der Waals equation of state *qualitatively* describes experimental observations, but not quite *quantitatively* so.

A similarly straightforward calculation, using the universal version of the van der Waals equation of state, Eq. (1.2), and demanding that $\partial p_r/\partial \rho_r = 0$, gives for the spinodal density as a function of the temperature

$$\rho_r^{spinodal} \sim 1 \pm \frac{2}{\sqrt{3}}\sqrt{1 - T_r}, \tag{1.4}$$

---

[3] A list of values of the van der Waals constant a and b for a large number of substances can be found on Wikipedia. See: https://en.wikipedia.org/wiki/Van_der_Waals_constants_(data_page).

[4] Arrows up or down, in expressions such as $x \uparrow\downarrow a$, mean that the value of $x = a(1 \mp \varepsilon)$ approaches that of a, with ε a positive number that is very much smaller than unity and the minus (plus) sign for the arrow up (down).

[5] The symbol $\sim$ signifies that the quantity on the right-hand side approaches asymptotically that on the left-hand side in the appropriate limit. Simply put, if $f(x) \sim g(x)$ as $x \to x_0$, implies that $f/g \to 1$ in that limit. Here, the horizontal arrow denotes that the value of the symbol on the left-had-side of it approaches that on the right-hand-side (from above or below).

for temperatures just below the critical temperature, so for conditions where $T_r \uparrow 1$, or, equivalently, $T \uparrow T_c$. (See also exercise 2 at the end of this chapter.) Here, the minus sign gives the spinodal density on the gas-phase side of the critical density, for which by definition $\rho_r = 1$, and the plus sign that on the liquid-phase side. By comparing Eqs. (1.3) and (1.4), we find that the spinodal region lies within the binodal region of the phase diagram, as is to be expected. See Fig. 1.2 (right).

In the region bounded by the spinodal, the *isothermal compressibility* $\beta_T \equiv -V^{-1} \times (\partial V / \partial p)_{N,T}$ of the homogeneous fluid is negative, noting that it can be rewritten as $\beta_T = (\rho \partial p / \partial \rho)^{-1}$. Consequently, a spontaneous density fluctuation in the gas will actually amplify itself! This then leads to the spontaneous and instantaneous emergence of a liquid phase in the gas or *vice versa*. As a matter of fact, the spinodal describes the maximum undercooling of a gas that can be achieved before it instantly condenses into a liquid, and the maximum overheating of a liquid before it instantly starts boiling. Hence, the spinodal is (in principle) a measurable quantity! A more practical method actually relies on measuring the amplitude of spontaneous density fluctuations, e.g., by radiation scattering methods.

In the region in the phase diagram bounded by the binodal and spinodal, the homogeneous gas (or liquid) is *metastable*, so is *locally* stable but not *globally* stable. If a homogeneous gas or liquid is suddenly brought ("quenched") in that metastable region, co-existence between a co-existing gas phase and a liquid phase is more stable than a homogeneous gas or liquid, but the isothermal compressibility of the fluid remains positive. Consequently, smallish density fluctuations die out spontaneously as the fluid relaxes back to a metastable state. Required are sufficiently large density fluctuations to initiate condensation or boiling, depending on whether we cross the binodal from the gas or the liquid side of the critical density.

Such large fluctuations are unlikely to happen because they cost free energy[6] and, because of that, there is a so-called *lag time* before condensation actually occurs via a mechanism known as *nucleation and growth* [3]. Nucleation refers to the process that only sufficiently large drops can grow, and that these drops can only form by climbing over a free energy barrier. Upon approaching the spinodal, the free energy barrier and the lag time go to zero. Beyond it, *any* spontaneous thermal fluctuation, no matter how small, drives macroscopic phase separation. This process is called *spinodal decomposition* [4].

Since the pioneering work of van der Waals, more sophisticated theories and computer simulations, rooted in statistical mechanical theory, have been put forward in order to increase our understanding of the structure and thermodynamics of simple gases, liquids and solids, and in particular phase transitions between them [5]. Indeed, phase diagrams such as shown in Fig. 1.2 can be understood on the basis of relatively simple pair-interaction potentials U, schematically represented in Fig. 1.4. The interaction potential $U(r)$ is presumed to be spherically symmetric and depends only on the distance r between the centres of mass of the two particles. The

---

[6] These fluctuations are expensive in terms of free energy, because they create a surface between gas and liquid and associated with that a surface free energy.

Fig. 1.4 Schematic of a spherically symmetric molecular interaction potential $U(r)$, where $r \equiv |\vec{r}|$ is the distance between the centres of mass of the particles and σ their size. See also the inset. The depth of the potential $-\epsilon$ measures the strength of the interaction. Indicated are where the steric repulsion and van der Waals attraction dominate. The position where the potential crosses zero is a measure for the size of a particle, σ.

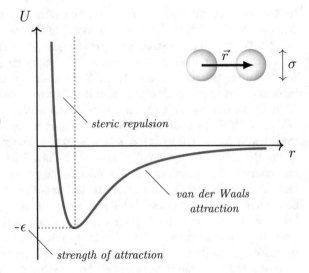

interaction potential ignores any atomic detail of the molecules, and is characterised by a steric repulsion at short inter-particle distances and by a van der Waals attraction at larger separations.

An in computer simulations often-used pair potential is the *Lennard-Jones potential* U_{LJ},

$$U_{LJ}(r) = 4\epsilon \left[\left(\frac{\sigma}{r} \right)^{12} - \left(\frac{\sigma}{r} \right)^{6} \right], \tag{1.5}$$

where σ is a measure for the size of the molecules and ϵ the strength of the Van der Waals interaction between them. The first term, proportional to r^{-12}, represents the steric repulsion between atoms and molecules, that is, the fact that no two particles can occupy the same volume in space.[7] The second term, proportional to r^{-6}, represents the effect of attractive van der Waals interactions between the particles, caused by permanent, induced, and quantum fluctuation-induced dipoles on them. The potential is equal to zero for $r = \sigma$, and has its minimum value of $U = -\epsilon$ at $r = 2^{1/6}\sigma \simeq 1.12\sigma$. This suggests that σ is a measure for the size of the particle (the so-called *Born radius*) and ϵ for the strength of the van der Waals attraction.

The two model parameters σ and ϵ play a similar rôle as the parameters a and b in the van der Waals equation of state, and have also been tabulated for many different compounds [6]. Indeed, they can be experimentally determined, e.g., by fitting to

---

[7] Overlap of electron clouds of atoms and molecules cause strong electrostatic repulsion between them. The fact that electrons are fermions, forbidding them to attain the same quantum numbers and hence occupy the same position, makes this repulsion even stronger.

the so-called *second virial co-efficient B* of the low-density (virial) expansion of the pressure for rarefied gases,

$$p = \rho k_B T (1 + B\rho + C\rho^2 + D\rho^3 + \cdots), \tag{1.6}$$

where C and D are the third and fourth virial co-efficients, and the dots represent a sum of terms of higher powers in the density. This expression was put forward in 1901 by Heike Kamerlingh-Onnes, also known as the discoverer of the phenomenon of superconductivity. If we truncate this expansion in powers of the density after the contribution from the second-virial term, $B\rho$, then this is called the *second-virial approximation* to the equation of state.

For particles interacting via a spherically symmetric potential, such as the Lennard-Jones potential, the second virial co-efficient can be written as[8]

$$B(T) = 2\pi \int_0^\infty dr\, r^2 [1 - \exp(-U(r)/k_B T)], \tag{1.7}$$

and follows from statistical mechanical theory of non-ideal gases. (See, e.g., the textbook of Richard Feynman [7] for a formal and also a less formal but physically more insightful derivation.) We return to the second virial co-efficient in Chap. 5, and in fact will obtain a generalised version of Eq. (1.7) using intuitive arguments and some basic thermodynamics.

For the Lennard-Jones gas, we expect $B_{LJ} = \sigma^3 f(\epsilon/k_B T)$, with $f(x)$ some complicated universal function of its argument x, here equal to the strength of the van der Waals attraction, ϵ, scaled to the thermal energy, $k_B T$. The reason is that we can use the size σ of the particle to make the integral dimensionless by a simple co-ordinate transformation. In exercise 1 at the end of this chapter, we arrive at the conclusion that for hypothetical model particles interacting via a Lennard-Jones potential, so-called *Lennard-Jonesium*, we must, to a good approximation have

$$B_{LJ} = \frac{2\pi}{3}\sigma^3 \left(1 - \frac{12}{5}\frac{\epsilon}{k_B T}\right), \tag{1.8}$$

whilst for the van der Waals equation of state, we have

$$B_{vdW} = b - \frac{a}{k_B T}. \tag{1.9}$$

By equating B_{vdW} and B_{LJ}, we can identify a direct correspondence between an interaction potential acting between the particles, and a phenomenological equation

---

[8] Note that we use the convention that $\int dx$ denotes an *integral operator* working on the variable x of functions $f(x)$ to the right-hand-side of it. Some readers might be more familiar with the convention $\int f(x)dx$, which we write as $\int dx f(x)$.

of state that was conceived explicitly with a particulate view of matter in mind. (See exercise 1!)

It stands to reason that for molecules that are not simple and not approximately spherical in shape, the interaction potential must be much more complex and for instance depend on their distance *as well as* on their relative orientation. This would in particular be important for the crystal structure, although a simple van der Waals theory of freezing that ignores all of this can be set up, as we show in exercise 3 of this chapter. From computer simulations, we know that model particles interacting via a Lennard-Jones potential exhibit at least two kinds of solid crystal phase, depending on temperature and pressure. These are known as the hexagonal closed packed crystal and face centred cubic crystal [8]. This is a far cry from what is seen experimentally for seemingly simple molecules such as methane (CH_4) or water (H_2O), of which respectively a half-dozen and dozen different crystal structures are currently known.

Perhaps not entirely surprisingly, more complex molecules may not only have complex crystal phases, but may in addition self-organise themselves in unusual aggregated states at temperatures in between where the "ordinary" liquid and crystal phases are stable. These additional phases are for that reason referred to as *mesophases*. We can distinguish two types of mesophase, known as *liquid crystals* and as *plastic crystals*. Plastic crystals are crystalline solids that exhibit positional order yet lack complete orientational order. The main body axis vectors of the particles point in that case exhibits randomness in at least one of the three so-called Euler angles that completely describe the orientation of an object in three-dimensional space. Liquid crystals are in some sense ordered fluids, in which one or more angular and/or positional degrees of freedom are *frozen-in*. "Frozen-in" here means that not all physically allowed orientations and positions are populated with equal probability. As liquid crystals are by definition *not* crystalline, this implies that no more than two *positional* degrees of freedom of the molecules can be frozen-in. In the next chapter, we will briefly touch upon liquid crystals with one or two dimensional crystalline order.

One of the purposes of the remainder of this book is to extend the concept of "order" and symmetry breaking from crystals to liquid crystals, that is, convey the idea that symmetry breaking not only refers to the appearance of positional order in three-dimensional space, but also to emergent orientational order that is connected to how particles are oriented in that space. We shall take one particular type of liquid crystal as an example for our discussions. This does not mean that other liquid-crystal phases, and there are very many of them, are not interesting, quite on the contrary, but that the ideas put forward carry over naturally to them.

As we are going to see in the next chapter, liquid crystals come in two "flavours" with very different thermodynamic driving forces that stabilise them. Luckily, although this is perhaps not always appreciated in the communities that study these two flavours of liquid crystal, the main theories describing them can derived from the same theoretical framework. Before going into an instructive albeit theoretically perhaps not necessarily rigorous derivation of that framework, let us first look into what liquid crystals are and in what way the two mentioned flavours of liquid crystal differ.

Further Reading A clear exposition of the workings of thermodynamics can be found in the textbook of C. J. Adkins [9]. For a derivation of the pressure within the second virial approximation, the reader is referred to the statistical mechanics textbook of Richard Feyman [7]. The classical textbook of David Tabor gives a lucid overview of the properties of gases, liquids, and solids, and other states of matter [10]. A slightly more technical discussion that includes complex fluids, polymers, and colloids, can be found in the book of Jean-Louis Barrat and Jean-Pierre Hansen [4].

Exercises

1. **Van der Waals and Lennard-Jones connected.** A direct comparison between the model parameters of the van der Waals equation of state, and those of the Lennard-Jones model potential, can be made by considering the second virial co-efficient, B. In this exercise, we verify the Eqs. (1.8) and (1.9) for the second virial co-efficients, make a connection between the van der Waals parameters and the Lennard-Jones parameters, and discuss the consequences.

   (a) Express the second virial co-efficient of the *van der Waals gas* in terms of the parameters a and b.[9] Taylor expand for this purpose the van der Waals equation of state for small values of $b\rho^2$, and retain only terms linear and quadratic in the density ρ. Recall that a Taylor expansion of a function $f(x)$ of a point $x + \delta$ near the continuous variable x obeys

$$f(x+\delta) = \sum_{n=0}^{\infty} \frac{1}{n!} f^{(n)}(x)\delta^n \approx f(x) + f'(x)\delta + \frac{1}{2}f''(x)\delta^2 + \cdots \quad (1.10)$$

   for $\delta \to 0$. Here, $f^{(n)}(x)$ denotes the nth derivative of the function with respect to x, evaluated at the point x, and f' and f'' the corresponding first and second derivative. Compare your findings with Eq. (1.9).

   (b) Calculate the second virial co-efficient of particles interacting via a Lennard-Jones potential using the expression of Eq. (1.7), and verify that Eq. (1.8) should be a good approximation. For this purpose, realise that for inter-particle distances r somewhat smaller than the particle diameter σ, $\exp[-U(r)/k_BT] \ll 1$ and can be neglected. For distances larger than this, $\exp[-U(r)/k_BT] \approx 1 - U(r)/k_BT$, in which case the van der Waals attraction term proportional to r^{-6} dominates the interaction potential.

---

[9] A van der Waals gas is the hypothetical gas that obeys the van der Waals equation of state.

(c) Relate the van der Waals parameters a and b to the Lennard-Jones parameters σ and ϵ. Discuss the temperature dependence of the second virial co-efficient, and what the consequences are for the pressure at low and high temperatures.

(d) The expression we found for the second virial co-efficient of the Lennard-Jones gas, Eq. (1.8), cannot be accurate at *extremely* high temperatures $T \gg \epsilon/k_B$. Discuss why this must be the case.

2. **Thermodynamics and the van der Waals equation of state.** According to the second law of thermodynamics, the isothermal compressibility $\beta \equiv -(\partial V/\partial p)_{N,T}/V$ of a gas must be a positive quantity under conditions of thermodynamic stability.

(a) Show that a positively valued isothermal compressibility implies that $\partial p/\partial \rho \geq 0$ and hence that $\partial^2 f/\partial \rho^2 \geq 0$, with $f = F/V$ the Helmholtz free energy F per unit volume V. The limit of thermodynamic stability, $\partial p/\partial \rho = 0$, is called the *spinodal*.

(b) Verify that for the van der Waals equation of state there must be combinations of T and ρ for which this condition is not held, and hence that the homogeneous gas or liquid cannot be thermodynamically stable under those conditions.

(c) Relate the critical density ρ_c and temperature T_c of the van der Waals equation of state in terms of the parameters a and b, using the fact that it is the maximum temperature of the spinodal.

(d) Verify that at the critical point the critical pressure p_c obeys the relation $p_c/\rho_c k_B T_c = 3/8$.

3. **Van der Waals theory of freezing.** Within the spirit of the van der Waals equation of state for gases and liquids, $(p + a\rho^2)(1 - b\rho) = \rho k_B T$, with a and b material-dependent parameters, a simple equation of state for the solid phase has been proposed based on a so-called cell model [11]. It takes the form $(p + \hat{a}\rho^2)(1 - (\hat{b}\rho)^{1/3}) = \rho k_B T$,[10] where $\hat{b}$ sets the maximum packing fraction in the crystal and $\hat{a}$ accounts for attractive interactions that in the solid phase are stronger due to more efficient packing of the particles. For particles interacting via strictly repulsive forces, $a = \hat{a} = 0$. Such particles are called *hard particles*.

(a) Using the thermodynamic relation $p = -\partial F/\partial V|_{N,T}$, verify that the Helmholtz free energy, F_{vdW}^f of a van der Waals fluid of hard particles obeys

$$\frac{F_{vdW}^f}{k_B T} = N\left[\ln(\rho\omega/e) - \ln(1 - \rho b)\right], \qquad (1.11)$$

---

[10] The one-third power of the density is the reciprocal of the average distance between the particles. See also [12].

with ω some temperature-dependent volume scale and $e = 2.718...$ Euler's number, whilst that of the solid of hard particles, F_{vdW}^s, obeys

$$\frac{F_{vdW}^s}{k_B T} = N \left[\ln (\rho \omega / e) - 3 \ln \left(1 - (\rho \hat{b})^{1/3} \right) \right]. \tag{1.12}$$

The terms logarithmic in ρ are the ideal gas terms, and Euler's number appears for historic reasons; see also Chap. 4. Hint: one can either verify that Eqs. (1.11) and (1.12) produce the given expressions for the equations of state, or, *vice versa*, use the equations of state to arrive at the Eqs. (1.11) and (1.12). The latter route requires a little cunning, namely the realisation that the Helmholtz free energy is an extensive variable and that the integration constant can only be a function of N and T.

(b) According to the equipartition of energy, the average kinetic energy per particle amounts to $3k_B T/2$ if the particles obey classical (Boltzmann) statistics. Show that this implies that $\omega \propto T^{-3/2}$. Hint: use the thermodynamic relation $U = \partial \beta F / \partial \beta$, where $\beta = 1/k_B T$.

(c) Explain why it is allowed to identify $\omega = b$ in both free energies, if we are interested in the difference between the free energies of the fluid and the crystal phases.

(d) Argue why co-existence between the fluid and crystal phases of the hard particles may be found by using the so-called *common tangent construction*. In the common tangent construction the free energy density F/V is plotted as a function of the density ρ, and the straight line that touches both free energy curves from below imposes equal pressures and equal chemical potentials. The common tangent construction is equivalent to the Maxwell construction mentioned in the main text.

For hard spheres, we can make a sensible *ansatz* for b and $\hat{b}$ in terms of the particle diameter σ. Let $\phi = \pi \sigma^3 \rho / 6$ be the volume fraction of the particles. The maximum volume fraction for randomly packed particles equals 0.64, whilst that for hexagonally closed packed particles equals 0.74. Hence, it seems reasonable to put $b\rho = \phi / 0.64$ and $\hat{b}\rho = \phi / 0.74$ as we expect the pressure to diverge for these volume fractions.

(e) Apply the common tangent construction using the van der Waals free energy densities of the fluid and crystal phases, and confirm that we expect fluid-crystal co-existence with respective volume fractions of approximately 0.63 and 0.71. More accurate calculations show that the freezing transition occurs at the somewhat lower volume fractions of 0.49 for the fluid and 0.55 for the crystal phase [4].

(f) The van der Waals model predicts an entropy-driven freezing transition for hard particles. Explain why this is called an entropy-driven transition and what this means for the entropy of the crystal phase relative to that of the fluid phase beyond some value of the density.

References

1. D. Chandler, *Introduction to modern statistical mechanics* (OUP, Oxford, 1987).
2. E. A. Guggenheim, *The Principle of Corresponding States*, J. Chem. Phys. **13** (1945), 253.
3. D. Kashchiev, *Nucleation, basic theory with applications* (Butterworth, Oxford, 2000).
4. J.-L. Barrat and J.-P. Hansen, *Basic concepts for simple and complex fluids* (CUP, Cambridge, 2003).
5. J.-P. Hansen and I.R. McDonald, *Theory of simple liquids - with applications to soft matter*, 4th edition (AP, Amsterdam, 2013).
6. F. Cuadros, I. Cachadiña, W. Ahumada, *Determination of Lennard-Jones interaction parameters using a new procedure*, Mech. Eng. **6** (1996), 319.
7. R. P. Feynman, *Statistical mechanics: a set of lectures* (CRC Press, Boca Raton, Florida, 2018).
8. H. Adidharma and S. P. Tan, *Accurate Monte Carlo simulations on FCC and HCP Lennard-Jones solids at very low temperatures and high reduced densities up to 1.30*, J. Chem. Phys. **145** (2016), 014503.
9. C. J. Adkins, *Equilibrium thermodynamics* (CUP, Cambridge, 1983).
10. D. Tabor, Gases, liquids and solids and other states of matter (CUP, Cambridge, 1991).
11. A. Daanoun, C. F. Tejero and M. Baus, *van der Waals theory for solids*, Phys. Rev. E **50** (1994), 2913.
12. H. N. W. Lekkerkerker and Remco Tuinier, *Colloids and the depletion interaction* (Springer, Dordrecht 2011).

Chapter 2
Liquid Crystals

Abstract In this chapter we discuss the concept of a broken symmetry in the context of positional and orientational degrees of freedom of molecules and particles, and introduce the simplest and most-studied liquid crystal, being the nematic liquid crystal. Using the concept of an order parameter, which may be seen as one (statistical) moment of the full orientational distribution function, we identify two distinct types of liquid crystal. These are known as thermotropic and lyotropic liquid crystals, and we highlight in what fundamental way they are different.

Liquid crystals possess properties common to both liquids and solids. Indeed, they flow like ordinary fluids do, but also respond elastically to certain types of mechanical deformation and in that sense resemble solids. Liquid crystals, in contrast to ordinary liquids, are *not* isotropic implying that their properties depend on the direction in which they are probed. While crystalline solids are characterised by long-range positional and orientational order, particles in liquid crystals may exhibit positional order in one or two dimensions, or in fact no positional order at all. In liquid crystals invariably one or more of the orientations of particles are "frozen-in", meaning that they do not attain all allowed values with equal probability. Having three positional degrees of freedom, and, on top of that, three angular degrees of freedom that fully describe the orientation of a particle in three-dimensional space, particles that are not spherical potentially exhibit a whole zoo of liquid crystal phases.

A small sample of liquid-crystalline phases is illustrated in Fig. 2.1. Shown are particles that point in random directions: this is the usual isotropic phase. If the particles align along a single common axis, they are in *nematic liquid crystal* phase. If they in addition order positionally along the common axis, i.e., form stacks of layers, then we are dealing with a *smectic A* phase. In the smectic A phase, there is no positional order within the layers that behave like weakly correlated two-dimensional fluids, while in the *columnar phase* there is two-dimensional positional order in the plane perpendicular to the preferred orientations of the particles but fluid-like order along it. One can perhaps view the columnar phase as a collection of hexagonally arranged, one-dimensional fluids. In the crystal phase, the particles

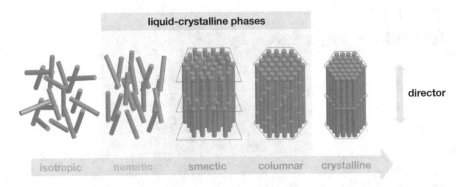

liquid-crystalline phases

director

isotropic   nematic   smectic   columnar   crystalline

Fig. 2.1 Selection of liquid-crystalline mesophases with increasing degree of order, in between the conventional isotropic fluid on the left to fully crystalline solid in the right. Indicated is also the director. See also the main text. Reproduced from [1], with the permission of AIP Publishing.

exhibit in-layer positional order and order across the layers. There are very many other liquid crystal states, such as the *biaxial nematic* in which two orientations are frozen-in, or the *chiral nematic* also known as *cholesteric* phase, where the preferred axis rotates with a fixed pitch in a direction perpendicular to it.

The simplest, and most comprehensively investigated liquid crystal is the nematic liquid crystal phase, or *nematic* for short. In the nematic phase, the molecules spontaneously and collectively align along a preferred direction, called the *director*. In other words, the nematic obeys uniaxial symmetry. The ground state of the director is uniform, so does not vary spatially. Any deformation of the director field costs free energy, and it is in this sense that nematics respond elastically. The director field may be deformed by means of an external field and/or by interaction of the director with the boundaries of the volume the nematic is held in. Actually, even though the ground state of a bulk nematic is uniform, spontaneous fluctuations turn out to become less costly in terms of free energy the larger the wave length of the fluctuation is. Such long-wave-length fluctuations are therefore thermally excited and in the end explain why nematics appear turbid. Indeed, director field fluctuations cause light to be scattered on account of the difference between the refractive indices parallel and perpendicular to the main optical axis of the material, which usually coincides with the director.

Not surprisingly, molecules that under appropriate conditions form a nematic phase are anisometric, and, broadly speaking, either elongated and rod-shaped, or flat and disk-like. The former form what are known as *calamitic nematics*, while the latter produce *discotic nematics*. A compound that is widely used in display technology applications, and that exhibits a nematic phase around room temperature, is 4-cyano-4'-pentylbiphenyl, also referred to as 5CB and shown in Fig. 2.2. Just like other compounds that form nematic phases, collectively known as

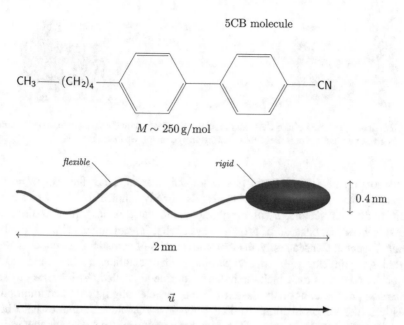

Fig. 2.2 Representation of a 5CB molecule focusing on the chemistry (top), on the physical chemistry (middle) and on the essence of the molecule in a (theoretical) physics model description, which is the main body axis vector $\vec{u}$ (bottom). The aromatic moiety on the right of the chemical structure formula is rigid, while the aliphatic tail on the left is flexible due to rotation about the chemical bonds between the carbon atoms.

nematogens[1], the molecular structure of 5CB is characterised by a rigid (aromatic) core attached to which is a flexible tail, often a short aliphatic (hydrocarbon) chain. The chemical structure of the molecule can be represented by a rigid, cigar-shaped particle with a flexible tail. The rigid body defines a unit body axis vector $\vec{u}$ with $|\vec{u}| \equiv 1$, and has a strong tendency to crystallise. The flexible tail introduces just enough disorder into the molecular structure to allow for the crystal phase to be destabilised in favour of the nematic phase, although that this only happens in a certain temperature range. Even though that molecules like 5CB are not symmetric, and resemble tadpoles, with the rigid portion of the molecule representing the head and the flexible part the tail, the nematic phase that they form do exhibit up-down symmetry: half the molecules point in the direction of the director and half point in the opposite direction.

For 5CB under conditions of atmospheric pressure, the sequence of phase transformations is (1) crystal → nematic liquid crystal at 18 °C, and (2) nematic liquid crystal → isotropic fluid at 35 °C. The isotropic fluid phase is equivalent to the usual liquid state. In the nematic phase, the main body axis vectors are more or less aligned along some preferred axis, which is the earlier-introduced director. The director is typically represented by a unit vector $\vec{n}$, implying that $|\vec{n}| \equiv 1$.

---

[1] More generally, particles that form liquid crystalline phases are referred to as mesogens.

Fig. 2.3 Isotropic liquid phase (left) and nematic liquid-crystalline phase (right). The common axis is given by the director $\vec{n}$. Because of inversion symmetry, we have $\vec{n} = -\vec{n}$.

See also Fig. 2.3. The nematic phase not only has uniaxial, that is, *cylindrical* symmetry but also *inversion* symmetry, as already alluded to.[2] This means the main body axis vectors, $\vec{u}$, of the particles are more or less aligned along the director $\vec{n}$, which, because of inversion symmetry, is sometimes called a "double-pointed" vector. Consequently, the thermodynamics of nematics does not change by inverting the director $\vec{n} \rightarrow -\vec{n}$, and particles with orientations $\vec{u}$ and $-\vec{u}$ are equally probable. Nematics for which the latter is not true are called *ferroelectric* or *polar nematics*. For these, not only the axes of the particles align, they in addition point mostly in the same direction. To date, only few polar nematic phases have been found experimentally, in spite of the fact that nematogenic compounds themselves typically are not inversion symmetric and hence polar.

Nematic liquid crystals have highly unusual physical properties because of their uniaxial symmetry. For instance, nematics are *birefringent*: The optical properties along the director are different from those perpendicular to that, which is why they find applications in opto-electronic technology, including displays and optical switches. Incidentally, in opto-electronic applications, different nematogens are typically mixed to optimise the isotropic-nematic transition temperature, response time, and so on, and nematics are in that case not strictly pure fluids. Actually, it is in practise very difficult to synthesise molecules of perfect purity, that is, without any contaminants. If in sufficiently low quantities, these contaminants do not strongly affect the properties of the liquid crystal, and the fluid can be treated as a single-component one.

Even as single-component fluids, nematics have a very complex flow behaviour depending on the flow direction and type, involving no fewer than five different "viscosities" (so-called *Leslie co-efficients*). As already mentioned, nematics respond elastically to distortions of the director field, the behaviour of which is described by five different elastic constants depending on the type of deformation. The main ones are called "splay", "twist" and "bend". See Fig. 2.4.

Finally, there are (in principle) three types of interfacial tension, depending on the attack angle of the director relative to the normal to the interface and a preferred axis in the plane of the interface, if that interface as a sense of direction and hence is not isotropic. In many opto-electronic applications, such as the twisted nematic cell in

---

[2] The symmetry group of nematics is $SO(3)$.

splay

twist

bend

Fig. 2.4 Nematic fluid responds elastically to different types of deformation of the director field. The main kinds of deformation of the director field are "splay" (left), "twist" (middle), and "bend" (right).

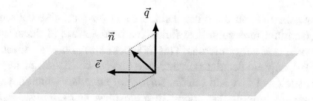

Fig. 2.5 Alignment of a director $\vec{n}$ to a solid surface with surface normal $\vec{q}$ and easy axis $\vec{e}$. In principle, the interfacial tension between the nematic and the surface depends on both the angle of the director with the surface director and the easy axis. For isotropic surfaces, the interfacial tension depends only on the angle between the director and the surface normal.

display technology, surfaces are treated such that the nematic develops a preference for aligning along some "easy" axis in the plane of the surface. See Fig. 2.5. That the surface tension of nematics depends on the angle of the director with the interfaces may cause nematic droplets (also known as *tactoids*) to attain a shape that is not spherical, but resembles a spindle or discus depending on the preferred angle of attack of the director field.

Although interesting, all of this is outside the scope of this book, which focuses entirely on molecular theories of why nematic phases arise in the first place. Molecular theories of nematics explicitly incorporate the notion of particles and interactions between them into their description. For this to work for the two types of liquid crystal that we distinguish, and that are commonly referred to as *lyotropic nematics* and *thermotropic nematics*, we need to consider in more detail what kind of physical system these are comprised of.

Thermotropic nematics are typically one-component fluids, that is, pure or nearly pure fluids of relatively low molecular weight materials, such as the example of 5CB discussed above. Although in practical application thermotropic nematics are often mixtures, they are often treated as pure fluids nonetheless. As a matter of fact, there are very many chemical compounds (and mixtures thereof) known that form nematic phases. Finally, there are also so-called main-chain and side-chain liquid crystal polymers comprised of chemically linked nematogen building blocks that in the molten state exhibit liquid crystal phases. See Fig. 2.6.

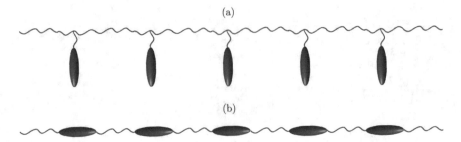

Fig. 2.6 Schematic representation of side-chain liquid-crystalline polymers (**a**) and main-chain liquid-crystalline polymers (**b**).

Lyotropic nematics, on the other hand, are rod- or plate-like colloidal particles finely dispersed in a molecular host fluid. The particles are in the size range from tens of nanometres to a few microns. This includes *chromonics*, a collective noun for self-assembled, supramolecular polymer-like objects consisting of dye molecules, *biopolymers* such as DNA and f-actin, *surfactant micelles*, which are supramolecular aggregates of amphipolar, soap-like molecules, and *colloidal particles* such as carbon nanotubes, graphene flakes, and clay particles. Of particular interest are rod-like viruses, such as *tobacco mosaic virus* (TMV) and *fd virus*, on account of their *monodisperse* character with strongly peaked length and width distributions [2, 3]. Both are seen as ideal model systems to study lyotropic nematics, not only because of their monodisperse character but also because of their relative ease to modify them chemically or genetically, and to produce in significant quantities [4]. Here, the term monodisperse refers on the one hand to the chemical makeup of the particles being identical and on the other that their shape and size are also essentially the same.

Particle-based theories that explain why molecules or colloidal particles spontaneously align along a common axis and form a nematic liquid crystal rely on the concept of an *orientational probability distribution function* or *orientational distribution function* for short. The orientational distribution function, $P(\vec{u})$, describes the probability density that the main body axis vector of a particle is in the angular region between $\vec{u}$ and $\vec{u} + d\vec{u}$. If the director $\vec{n}$ is defined along the z-axis of a Cartesian co-ordinate system, then

$$\vec{u} = (\sin\theta\cos\phi, \sin\theta\sin\phi, \cos\theta)^T \qquad (2.1)$$

can be described in the usual polar and azimuthal angles, with $\theta \in [0, \pi]$ and $\phi \in [0, 2\pi]$ radians. Because nematics obey cylindrical symmetry, we have $P(\vec{u}) = P(\theta, \phi) = P(\theta)$. On account of the inversion symmetry of the nematic, we in addition have $P(\vec{u}) = P(-\vec{u})$, which translates to $P(\theta) = P(\pi - \theta)$. See Fig. 2.7. Probability distribution functions are normalised by definition, so $\int d\vec{u}\, P(\vec{u}) = \int_0^\pi d\theta \sin\theta \int_0^{2\pi} d\phi\, P(\theta) = 1$. For an isotropic phase, $P(\vec{u}) = P(\theta, \phi) = P(\theta) = 1/4\pi$ because $\int d\vec{u} = 4\pi$.

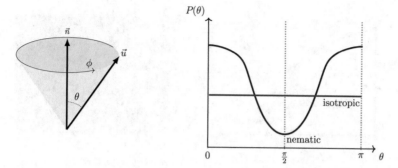

Fig. 2.7 Left: the main body axis vector $\vec{u}$ of a nematogen makes an angle θ with the director $\vec{n}$. The distribution function of the particles in the nematic phase depends only on this angle, not on the azimuthal angle ϕ. Right: the orientational distribution function, $P(\vec{u}) = P(\theta)$, of a nematic. Because the nematic has cylindrical and inversion symmetry, the distribution function must be a symmetric function about $\theta = \pi/2$, indicated by the dashed line in the middle. It is symmetric about $\theta = 0$ and $\theta = \pi$ too.

It turns out useful for our discussion to not only consider the orientational distribution function $P(\vec{n} \cdot \vec{u})$, where $\vec{n} \cdot \vec{u} \equiv \cos\theta$ and $\cdot$ the usual dot product, but also one of its (statistical) moments known as the *scalar nematic order parameter*. The scalar nematic order parameter denoted S, not to be confused with the entropy, and defined as[3]

$$S \equiv \langle P_2(\vec{n} \cdot \vec{u}) \rangle \equiv \left\langle \frac{3}{2}(\vec{n} \cdot \vec{u})^2 - \frac{1}{2} \right\rangle = \left\langle \frac{3}{2}\cos^2\theta - \frac{1}{2} \right\rangle \tag{2.2}$$

where

$$\langle \cdots \rangle \equiv \int d\vec{u}\, P(\vec{u})(\cdots) = \int_{-1}^{1} d\cos\theta \int_{0}^{2\pi} d\phi\, P(\theta, \phi)(\cdots) \tag{2.3}$$

denotes an orientational average, and $P_2(x) \equiv (3x^2 - 1)/2$ is the *second Legendre polynomial* [5].

For perfectly aligned particles we can easily verify that $S = 1$, while for randomly oriented particles we have $S = 0$. If the particles are *disaligned*, and point perpendicular to the director, we have $S = -0.5$. (See also exercise 2 below.) Typical values obtained experimentally by nuclear magnetic resonance, optical and spectroscopic methods for the nematic order parameter are $S \approx 0.3$–0.5 for thermotropic nematics and $S \approx 0.6$–0.8 for lyotropic nematics, at least for conditions near the melting transition to the isotropic phase. The order parameter increases in value with *decreasing* temperature for thermotropics, and with *increasing* concentration for lyotropics.

---

[3] Note that all odd moments vanish because of inversion symmetry.

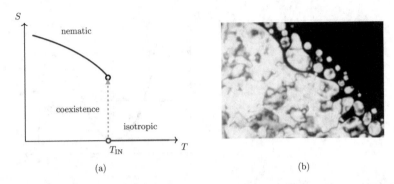

(a)   (b)

Fig. 2.8 (**a**) Nematic order parameter S of a thermotropic liquid crystal as a function of the temperature T. T_{IN} is the isotropic-nematic transition temperature at which co-existence between isotropic and nematic phases is possible. (**b**) Polarisation microscopic image of 5CB at the isotropic-nematic transition temperature. Black: isotropic phase, light coloured: co-existing nematic phase. Photographic image: Liquid Crystal Group, Institute of Physics, Jagiellonian University, Krakow, Poland. Reproduced with permission.

For thermotropics, the phase transition from an isotropic fluid to a nematic fluid is driven by enthalpy, i.e., by attractive (van der Waals) interactions between the molecules the strength of which depends on the angle between them [6]. These have to be sufficiently strong to overcome the loss of orientational entropy upon alignment. This implies that for thermotropics the *temperature* is the relevant thermodynamic control variable. Indeed, if we plot the order parameter S as a function of temperature T for a thermotropic nematic, we find that at the isotropic-nematic transition temperature, T_{IN}, which is also known as the *clearing temperature*,[4] the order parameter jumps from $S = 0$ to a non-zero value of $S \approx 0.3$–0.4. In other words, the isotropic-nematic transition is *discontinuous*, and therefore of the *first order*. Cooling an isotropic fluid of nematogens to the clearing temperature T_{IN} allows for the co-existence of isotropic and nematic phases, provided that not of all of the latent heat has been extracted from the fluid. See Fig. 2.8.

Lyotropic nematic liquid crystal transitions are driven, in essence, by the competition between two different types of entropy: translational entropy and orientational entropy. The reason is that aligned particles pack more efficiently than randomly oriented ones do. This increases their free volume albeit that this comes at the expense of an angular "confinement" that has associated with it a loss of orientational entropy. Because of this, varying the temperature does not have a large impact on lyotropic systems, and is not a useful thermodynamic control variable for this kind of system. Indeed, lyotropic nematics turn out to often behave approximately athermally, that is to say that they only respond weakly to

---

[4] As already alluded to, the nematic phase is turbid due to director fluctuations. The isotropic phase does not have a (fluctuating) director field and hence is not turbid. Melting a nematic leads to the emergence of a "clear", isotropic fluid.

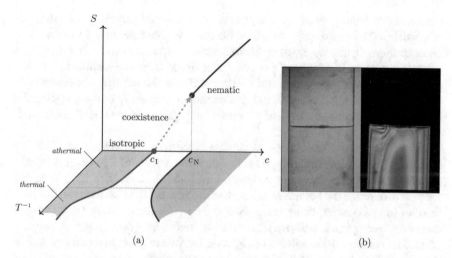

(a)   (b)

Fig. 2.9 (a) Order parameter of lyotropic liquid crystals as a function of the concentration c. The binodals, given by the concentrations c_I and c_N in the co-existing isotropic and nematic phases, are plotted as a function of the reciprocal temperature T^{-1} (drawn lines). In the athermal region energy plays a minor rôle and entropy dominates. In the thermal region, energy plays an important rôle and may involve gelation and solubility-limit effects that widen the phase gap. (b) Crucible containing an aqueous dispersion of tobacco mosaic virus, viewed directly (left) and through crossed polarisers (right), highlighting the isotropic top phase and the birefringent character of the nematic bottom phase. Photographic image reproduced from [7], with the permission of AIP Publishing.

temperature changes, at least in some range of temperatures.[5] In that temperature range, the thermodynamic state of the fluid is dominated by entropy and enthalpy plays only a subdominant rôle. Translational entropy depends on the availability of *free volume*, which is the volume *not* taken up by the particles present in the volume. As we shall see in Chap. 5, the free volume depends on how many particles there are in a given volume as well as their degree of alignment. Hence, the concentration must in this case be the relevant experimental control variable, and to a much lesser extent the temperature. See Fig. 2.9, showing schematically how the order parameter S depends on the particle concentration and temperature.

The figure shows that with increasing concentration the order parameter jumps from zero to some value between, say, 0.5–0.8, depending on the ionic strength, solvent quality, and so on, of the dispersion of particles. Of course, for sufficiently low temperatures, enthalpy can become important again. In that case the particles drop out of solution or, alternatively, form an arrested, gel-like state. Notice that under conditions of co-existence, not only the value that the nematic order parameter

---

[5] If the particles are self-assembled, such as is the case in chromonics and cylindrical micelles, the mean particle length may be affected by the temperature. The same is true for the impact of a molecular bending stiffness. See Chap. 7.

takes on in the isotropic and nematic phases are different, but also the concentrations of particles. The *phase gap*, which in this case is the difference in the particle concentrations in the co-existing isotropic and nematic phases, can range from a few per cent to hundreds of per cent. This depends on the architecture of the particles, e.g., their aspect ratio and bending stiffness, the solution conditions such as the acidity and ionic strength, but also on the *polydispersity* of the particles: the distribution of lengths or widths or the nematogens present in the host (background) fluid.

In both lyotropic and thermotropic liquid crystals, the transition from the isotropic (I) to the nematic (N) phase is discontinuous, so, of first order. Literally a plethora of van der Waals-type, lattice-based and field theories have been put forward to describe the I-N transition. Aim of this monograph is to focus on the main theories in this context, being Maier–Saupe theory for thermotropics and Onsager theory for lyotropics, and derive these from an overarching description. As already alluded to, our exposition will not be rigorous, but focuses instead on understanding the underlying physical principles and does not require a deep understanding of statistical mechanics even though we will make use of it.

Further Reading Readers interested in the fascinating history of liquid crystal science and technology are recommended to consult the book of David Dunmur and Tim Slukin [8]. A classic and accessible textbook on the properties and basic theories of many kinds of liquid crystal is that of Sivaramakrishna Chandrasekhar [9]. The standard but more technical tome on the physics of liquid crystals is that of Jacques Prost and Piere-Gilles de Gennes [10]. Recommended in this context is also the soft matter textbook by Maurice Kleman and Oleg Lavrentovich, discussing a wide range of aspects of the physics of liquid crystals [11]. A more recent review of a modern take on liquid crystals as dispersants of nanoparticles can be found in the two-volume book edited by Jan Lagerwall and Chiusi Scalia [12].

Exercises

1. **Cooling down thermotropics.** Matter cools down if heat is extracted from it. How much heat has to be extracted to lower the temperature by one degree depends on the value of the isobaric heat capacity C_p [J K^{-1}]. The heat capacity depends on the amount of materials and is only weakly temperature dependent, except very near the clearing temperature, where it practically diverges.[6] Consider some volume of an isotropic fluid of a nematogen at temperature T [K] above the clearing temperature T_{IN} [K], so $T > T_{IN}$. At time zero, we extract with the aid of a cooling element energy at some fixed rate R [J s^{-1}].

---

[6] The volume has to be sufficiently large and the cooling process has to be sufficiently slow.

(a) Sketch the temperature T of the fluid as a function of the amount of heat $H = Rt$ [J] extracted in a certain amount of time t [s], indicate the clearing temperature T_{IN} and the latent heat H_{IN}, and explain the shape of the curve.

(b) Explain how this allows us to pinpoint the clearing temperature with great accuracy, at least if we cool sufficiently slowly, that is, if the rate R is small.

(c) Sketch how the volumes of the co-existing isotropic and nematic phases relative to the overall volume are related to the amount of energy H extracted. Ignore the impact of thermal compression upon cooling down, which is anyway very small.

2. **Distribution functions and order parameters.** The orientational distribution function $P(\vec{u}) = P(\vec{u} \cdot \vec{n})$ that describes the particles in the isotropic and nematic phases is a function only of the polar angle θ between the main body axis vector $\vec{u}$ of a nematogen and the preferred axis or director $\vec{n}$, taken to be the z-axis in a Cartesian co-ordinate system. In the isotropic phase, there is no director, but we can take any arbitrary direction to act as director.

(a) Argue why the first moment of the distribution function, $\langle \vec{n} \cdot \vec{u} \rangle$, must be zero in both the isotropic and the nematic phase.

(b) Presume the nematogen has a permanent dipole moment, $\vec{p}$. Discuss how the orientation of $\vec{p}$ relative to the main body axis vector $\vec{u}$ determines whether such a molecule aligns with an externally applied electric field $\vec{E}$ or at some other angle to that field.

(c) Verify that the value of the scalar nematic order parameter, $S \equiv \langle \frac{3}{2} (\vec{n} \cdot \vec{u})^2 - \frac{1}{2} \rangle$, attains the value of unity for perfectly aligned particles, of zero for particles that are randomly oriented, and negative one-half if perfectly disordered, that is, aligned perpendicular to the director.

(d) Discuss what kind of scalar nematic order parameter one would define in two rather than three spatial dimensions, if the requirement is that for perfectly aligned particles the value has to be unity, and naught for isotropically distributed ones.

3. **Van der Waals picture of nematics.** Let us see what needs to be done to extend the van der Waals theory of the condensation of gases nematic fluids to the isotropic-nematic transition of thermotropic nematics. We start with the Helmholtz free energy F_{vdW} of the van der Waals fluid discussed in Chap. 1, exercise 3. If we include the contribution from attractive interactions, we have

$$F_{vdW} = F_{id} + F_{exc}, \tag{2.4}$$

where

$$F_{id} = Nk_B T \ln (\rho \omega / e) \tag{2.5}$$

is, as we shall see in Chap. 3, the Helmholtz free energy of an ideal gas with ω for our purposes an unimportant quantum-mechanical volume scale, e Euler's number, and

$$F_{\text{exc}} = -Nk_B T \ln(1 - \rho b) - aN\rho \qquad (2.6)$$

an excess free energy that accounts for the interactions between the particles. Here, $\rho = N/V$ is the density of the fluid, with N the number of particles and V the volume, and $k_B T$ the usual thermal energy, with k_B Boltzmann's constant and T the absolute temperature. We recall that a measures the strength of attractive interactions between the particles and that b accounts for the repulsive interactions.

(a) Presume for simplicity that a and b do not depend on the temperature. Using the thermodynamic relation for the Helmholtz free energy of the fluid, $F = E - ST$, with E the internal energy and S the entropy, identify which parts of the *excess* free energy, Eq. (2.6), represent entropy and which energy. (Here, the symbol S should not be confused with that of the scalar nematic order parameter!) Does the entropy go up or down with increasing density of particles? Explain.

(b) Argue why for elongated particles we should expect a to *increase* and b to *decrease* with *increasing* degree of orientational order.

(c) *Phenomenologically*, we could model the dependence of the van der Waals parameters on the nematic order parameter S by writing $a(S) = a_0 + a_2 S^2 + \cdots$ and $b(S) = b_0 - b_2 S^2 + \cdots$ with $a_0, a_2, b_0, b_2 > 0$ phenomenological parameters [13]. Why do we expect the corresponding corrections of the parameters a and b to depend on the square of the order parameter, rather than a linear power?

(d) If b decreases with increasing degree of orientational order, the associated change of the entropy of the molecules must increase according to Eq. (2.6). What kind of entropy does this represent?

(e) What kind of entropy do you expect to *decrease* with increasing degree of order? Apparently, we did not account for this kind of entropy loss in our simplistic van der Waals theory. How could we *phenomenologically* correct for this?

References

1. B. de Braaf, M. Oshima Menegon, S. Paquay and P. van der Schoot, *Self-organisation of semi-flexible rod-like particles*, J. Chem. Phys. **147** (2017), 244901.
2. S. Fraden, *Phase transitions in colloidal suspensions of virus particles*, In "Observation, Prediction, and Simulation of Phase Transitions in Complex Fluids", M. Baus, L. F. Rull, and J. P. Ryckaert, Eds. (Kluwer Academic Publishers, Dordrecht, 1995).

3. M. P. Lettinga, *Viruses as colloidal model systems*, in: Physics meets biology, from soft matter to cell biology, C11, Eds. G. Gommper, U. B. Kaupp, J. K. G. Dhont, D. Richter and R.G. Winkler (Forschungszentrum Jülich GmbH, 2004).
4. E. Grelet, *De virus comme système modèle de cristaux liquides*, L'act. Chem. **347** (2010), 20.
5. M. Stone and P. Goldbart, *Mathematics for physics* (CUP, Cambridge, 2000).
6. V. A. Parsegian, *Van der Waals forces: a handbook for biologists, chemists, engineers, and physicists* (CUP, Cambridge, 2006).
7. Z. Dogic, P. Sharma, M. Zakhary, *Hypercomplex liquid crystals*, Ann. Rev. Condens. Matter Phys. **5** (2014), 137-157.
8. D. Dunmur and T. Sluckin, *Soap, and flat-screen TVs* (OUP, Oxford, 2011).
9. S. Chandrasekhar, *Liquid crystals*, 2nd edition (CUP, Cambridge, 1992).
10. P. G. de Gennes and J. Prost, *The Physics of Liquid Crystals*, Second Edition (OUP, Oxford, 1993).
11. M. Kleman and O. D. Lavrentovich (Springer, New York, 2003).
12. J. P. F. Lagerwall and G. Scalia, eds., *Liquid Crystals with Nano and Microparticles*, vol. I and II, (World Scientific, New Jersey, 2017).
13. G. Vertogen and W. H. de Jeu, *Thermotropic liquid crystals, fundamentals* (Springer, Berlin, 1988).

Chapter 3
The Groundwork

Abstract Basing ourselves on quite generic arguments, we present a free energy that acts as starting point for describing thermotropic and lyotropic nematic liquid crystals. This free energy is a function of the orientational distribution function of the nematogens and the molecular interaction field each nematogen experiences from all other nematogens in the fluid. The latter depends on the one hand on the orientational distribution of the particles and on the other hand on their local arrangement in the form of a function that describes the spatial correlation in space between pairs of particle.

Our aim is to derive, combining (1) intuitive arguments, (2) common sense, and (3) basic thermodynamics, a molecular theory that is sufficiently general to act the starting point for describing the isotropic-to-nematic transition in thermotropic systems and that in lyotropic systems. This will highlight the similarities between Maier–Saupe- and Onsager-type theories and show that these are two sides of the same coin.

We start by defining a Helmholtz free energy, F [J], and presume it to be some function $F[\rho(\vec{r}, \vec{u})]$ of the number density $\rho(\vec{r}, \vec{u})$ [m^{-3}] of particles with orientation, $\vec{u}$, and position of the centre of mass, $\vec{r}$. Because nematics are spatially uniform fluids, we have $\rho(\vec{r}, \vec{u}) = \rho(\vec{u}) \equiv \rho P(\vec{u})$ with $\rho = N/V$ [m^{-3}] the number density of the N particles in the system volume V [m^3], and $P(\vec{u})$ the earlier-introduced angular distribution function. We recall that this quantity is normalised, see Chap. 2.

In the theoretical literature, $F[\rho(\vec{r}, \vec{u})]$ is called a density *functional* rather than a function, as it is a function of the density distribution $\rho(\vec{r}, \vec{u})$, which itself is a function of position $\vec{r}$ and orientation $\vec{u}$ of a test particle, and assigns a number to that quantity for every pair of values of $\vec{r}$ and $\vec{u}$. To stress the distinction between functions and functionals, so between "functions of variables" and "functions of functions," we adorn the latter with the square brackets and the former with round brackets. A deep theorem from statistical mechanics now posits that the free energy F is a unique functional of the equilibrium density distribution for any given interaction potential acting between the particles [1]. The trouble now is that we

P. van der Schoot, *Molecular Theory of Nematic (and Other) Liquid Crystals*,
SpringerBriefs in Physics, https://doi.org/10.1007/978-3-030-99862-2_3

do not know beforehand the form that this unique functional has for our specific problem.

To make headway and set up the theory, we follow the following pragmatic recipe: First, we identify the ideal gas or ideal solution free energy, F_{id}. Second, we make use of what is known as the Gibbs entropy to formulate an expression for the contribution of the orientational entropy to the free energy, F_{or}. Third, we account for interactions between the particles through an excess free energy, denoted F_{exc}. Our estimate for the total free energy is then the sum of these three contributions:

$$F = F_{id} + F_{or} + F_{exc}. \tag{3.1}$$

Let us discuss the three contributions to the free energy in more detail and start with the first contribution for a thermotropic liquid crystal, for which the ideal gas is a good starting point. For a closed system, i.e., one defined by the number of particles N, volume V, and absolute temperature T,[1] we know that the pressure p [N m^{-2}] obeys the ideal gas law $\lim_{\rho \to 0} p = -(\partial F / \partial V)_{N,T} = \rho k_B T$ in the low-density limit. (Recall Chap. 2!) Here, k_B [J K^{-1}] denotes, as usual, Boltzmann's constant. By isothermal integration over the volume and demanding that the free energy must be an extensive quantity, we are able to obtain the ideal gas contribution to the free energy, $F_{id} = -\int dV P|_{N,T} = N k_B T \ln(\rho \omega / e)$ with $\omega = \omega(T)$ a temperature-dependent volume scale and e Euler's number. For details of the derivation, see exercise 3 of Chap. 2.

The volume scale can be calculated exactly by taking the classical limit in the quantum-statistical theory of non-interacting point particles, from which Euler's number e also emerges naturally. This identifies $\omega = \Lambda^3$ with $\Lambda = \sqrt{h^2 / 2\pi m k_B T}$ [m] the *thermal de Broglie wavelength* of the particle. Here, m [kg] denotes the mass of the particle and h [J s^{-1}] is the Planck's constant. The thermal wavelength is a measure for how spread out the wave packet of the point particle is. In the limit where classical (Boltzmann) statistics holds, the mean distance between the particles should far exceed the thermal wavelength. For molecules, this for all intents and purposes is always the case at room temperature, even in dense fluid or solid phases.

In lyotropic nematics, we are dealing with a carrier fluid and, dispersed in it, colloidal particles. The appropriate reference state is then not an ideal gas but an ideal dispersion of non-interacting particles. In this case, the relevant quantity is not the pressure p, but a quantity known as the *osmotic pressure* Π. It is defined as $p = p_S + \Pi$, where p_S is the pressure of the pure solvent in a hypothetical reservoir that the solvent in the dispersion is in chemical and thermal equilibrium with. For the half-open (N, V, T, μ_S) system of interest,[2] where μ_S [J] is the

---

[1] Not to be confused with an *isolated system* for which the number of particles N, the volume V, and the internal energy E are fixed.

[2] In statistical mechanics, this is called a semi-grand canonical ensemble. Here, the solvent molecules can be exchanged between system and reservoir, but the solute particles cannot as they

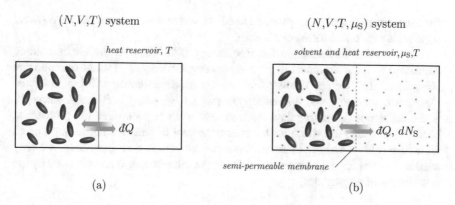

(N,V,T) system (N,V,T,μ_S) system

heat reservoir, T *solvent and heat reservoir,* μ_S,T

dQ $dQ,\, dN_S$

semi-permeable membrane

(a) (b)

Fig. 3.1 (**a**) A closed or (N, V, T) system in thermal equilibrium with a heat reservoir at temperature T. (**b**) Half-open or (N, V, T, μ_S) system in thermal and chemical equilibrium with a heat and solvent reservoir, at absolute temperature T and solvent chemical potential μ_S. Heat transfer between system and reservoir is indicated by dQ and particle transfer by dN_S.

chemical potential of the solvent in the reservoir, and the osmotic pressure obeys $\lim_{\rho \to 0} \Pi = -(\partial F/\partial V)_{N,T,\mu_S} = \rho k_B T$ according to the van 't Hoff law for ideal solutions. See also Fig. 3.1 for an explanation of the fundamental difference between a closed and half-open system. For an ideal solution, we obtain what in essence is the same result as that we obtained for an ideal gas, that is, $F_{id} = -\int dV \Pi|_{N,T,\mu_S} = N k_B T \ln(\rho\omega/e)$. The only difference is that the microscopic volume $\omega = \omega(T, \mu_S)$ [m^3] now depends not only on the temperature, T, but also on the chemical potential, μ_S, of the solvent.

An expression for the microscopic volume ω is not straightforwardly identified, if only because we do not to seek to explicitly model the solvent medium. The solvent medium itself is a highly non-ideal and strongly interacting collection of particles that in addition interact with the colloids or molecules dispersed in it. The implication is that ω must account for the amount of work done in order to insert a colloidal particle or molecule into the solvent medium. Hence, it depends not only on the chemical and structural makeup of the solvent molecules but also that of the particles dispersed in it.

Fortunately, the microscopic volume only acts to shift the free energy and hence may be viewed as a reference free energy for the ideal gas or ideal dispersion. Because of that, it has no impact on any phase behaviour. For this reason, we write the same expression

$$F_{id} = N k_B T \ln(\rho\omega/e), \tag{3.2}$$

remain confined to the volume V. In practice, this can be realised by separating dispersion and solvent reservoir by a semi-permeable membrane with pores sufficiently large to allow the solvent molecules to pass, but too small for the colloidal particles to do the same.

for thermotropics *and* lyotropics and need not worry about ω as it has no physical significance for constructing our theory.

The second contribution to the free energy is that due to the Gibbs entropy associated with the alignment of the particles that we denote S_{or}, where the subscript reminds us that we are dealing with an *orientational* entropy. From statistical mechanics, we know that this entropy obeys $S_{\text{or}} = -k_{\text{B}} \sum_\nu P_\nu \ln P_\nu$, where P_ν is the probability of a so-called *microstate* ν of the system. For the problem at hand, a microstate is a distinguishable arrangement of particles in the space of all allowed positions and orientations. Since the nematic is a uniform fluid, we only need to account for the orientations. Hence, the microscopic state is defined by the orientations of all particles,

$$P_\nu = P(\vec{u}_1, ..., \vec{u}_N) = P(\vec{u}_1)P(\vec{u}_2) \cdots P(\vec{u}_N) = \prod_{i=1}^{N} P(\vec{u}_i) \tag{3.3}$$

at least within a classical description of the N particles, where they can be numbered $i = 1, ..., N$. Implicit in Eq. (3.3) is also a mean-field approximation, where the orientations $\vec{u}_i$ of the individual particles are presumed independent of each other. Within this approximation, the multi-particle probability distribution function becomes the product of the one-particle distribution functions. In other words, the multi-particle distribution function *factorises*.

We next realise that $\sum_\nu = \prod_{i=1}^{N} \int d\vec{u}_i$, and make use of (1) the realisation that $\ln P_\nu = \ln \prod_{i=1}^{N} P(\vec{u}_i) = \sum_{i=1}^{N} \ln P(\vec{u}_i)$, (2) the presumed independence of $P(\vec{u}_i)$ from $P(\vec{u}_{j \neq i})$, and (3) of the property of normalisation of all distribution functions $P(\vec{u}_i)$, $\int d\vec{u}_i P(\vec{u}_i) = 1$. This gives for the orientational free energy, $F_{\text{or}} = -T S_{\text{or}}$, the following expression:

$$F_{\text{or}} = k_{\text{B}} T N \int d\vec{u} P(\vec{u}) \ln P(\vec{u}), \tag{3.4}$$

where $P(\vec{u})$ denotes the orientational distribution function of an arbitrary test particle, all particles being identical. Notice that we could also have written Eq. (3.4) straight away because of the additivity property of the entropy of N independent particles: the entropy of a collection of N non-interacting particles is the sum of the entropy of the N individual particles.

Finally, we need to address the issue of the excess free energy. This free energy represents all relevant additional contributions not taken into account by the ideal free energy and the orientational free energy. The main contribution comes from interactions between the nematogens. As these particles are chemically complex, and in the case of a lyotropic system also involves contributions from the presence of a solvent, we need to simplify the description by introducing the concept of *potentials of mean force*. In potentials of mean force, irrelevant degrees of freedom

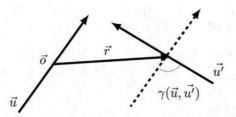

Fig. 3.2 The pair potential $U(\vec{r}, \vec{u}, \vec{u}')$ is a complicated function of the relative position $\vec{r}$ and the orientations $\vec{u}$ and $\vec{u}'$ of two test particles. For a given angle $\gamma(\vec{u}, \vec{u}')$ between the main body axis vectors of the two test particles, we expect a Lennard–Jones type of interaction potential characterised by a short-range repulsion and a long-range van der Waals attraction. The strength of the attraction depends on the relative orientation of the particles. See Fig. 3.3.

are, in a way, "integrated out."[3] We already encountered this, when we ignored the solvent molecules in our description of the free energy of an ideal dispersion.

In addition, atomic details in the interaction between molecules are effectively lost in the actual interaction potential even for quite small distances due to what perhaps may be called *self-averaging*: the interaction potential is the sum (or average) of the interaction potentials between all atoms of both particles as these are relatively long-ranged. In practice, this means that we presume the atomic details of the interaction between nematogens to be hidden in the phenomenological potentials that we make use of.

For simplicity, we further presume that the only relevant degrees of freedom are the orientations of the two particles $\vec{u}$ and $\vec{u}'$ and the position vector $\vec{r}$ connecting the centre of mass of one particle to that of the other. See Fig. 3.2. So, we expect the pair potential $U = U(\vec{r}, \vec{u}, \vec{u}')$ to be a function only of $\vec{r}$, $\vec{u}$, and $\vec{u}'$. For any given angle $\gamma = \gamma(\vec{u}, \vec{u}')$ between the two main body axis vectors $\vec{u}$ and $\vec{u}'$, the interaction potential should arguably be Lennard–Jones-like, as shown in Fig. 3.3. Note that the interaction potential need not be symmetric for angles $0 \leq \gamma < \pi/2$ and $\pi/2 < \gamma \leq \pi$ because the particles need not be inversion (or head–tail) symmetric.

The overall potential energy U_{pot} of a collection of N nematogens in a volume V depends on the instantaneous positions and orientations of all the particles in that volume. To calculate the expectation value of this quantity, which is the thermodynamic value of the potential energy, we rely on two approximations. First, we presume that the interactions between the particles are pair-wise additive. Because in reality this is generally not the case, we need to "renormalise" the interaction in order to get a description that somehow accounts for this. As a result, the interaction potential can no longer be predicted, e.g., from quantum mechanics, and has to be seen as a phenomenological potential. Second, we invoke the mean-

---

[3] Strictly speaking, these degrees of freedom are literally integrated out in the partition function of the system, leaving only the degrees of freedom of the interesting particles.

Fig. 3.3 Absolute value of the interaction potential $|U|$ between two nematogens positioned at some distance from each other need not be symmetrical for parallel configurations (with angle $\gamma = 0$) and for antiparallel configurations (with angle $\gamma = \pi$). See also Fig. 3.2.

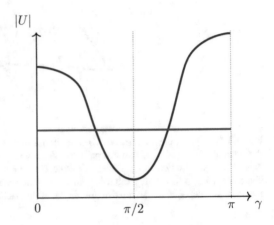

field approximation, which in fact we already did in formulating the orientational free energy.

Within these approximations, we write

$$U_{\text{pot}} = \frac{1}{2}N \int d\vec{u}\, P(\vec{u})U_{\text{mol}}(\vec{u}),\qquad(3.5)$$

where the factor $\frac{1}{2}$ corrects for double counting, and $U_{\text{mol}}(\vec{u})$ is the "molecular field" each particle experiences from the presence of all the other particles in the volume. The molecular field obeys

$$U_{\text{mol}}(\vec{u}) = \int d\vec{r} \int d\vec{u}'\, \rho(\vec{r}, \vec{u}'|\vec{0}, \vec{u})U(\vec{r}, \vec{u}, \vec{u}'),\qquad(3.6)$$

with $\rho(\vec{r}, \vec{u}'|\vec{0}, \vec{u})$ the number density of particles at position $\vec{r}$ with orientation $\vec{u}'$, given that there is a test or reference particle with orientation $\vec{u}$ at the origin $\vec{0}$ of our co-ordinate system. Hence, $\rho(\vec{r}, \vec{u}'|\vec{0}, \vec{u})$ denotes a *conditional* number density. Equation (3.6) literally sums the interaction energy a test particle experiences from all other particles in the volume.

The conditional density $\rho(\vec{r}, \vec{u}'|\vec{0}, \vec{u})$ is usually converted into a so-called *pair correlation function*, defined as

$$\rho(\vec{r}, \vec{u}'|\vec{0}, \vec{u}) \equiv \rho(\vec{u}')g(\vec{r}, \vec{u}, \vec{u}') = \rho P(\vec{u}')g(\vec{r}, \vec{u}, \vec{u}'),\qquad(3.7)$$

where we have made use of the fact that far from the test particle located at the origin, we expect $\rho(\vec{r}, \vec{u}'|\vec{0}, \vec{u}) \to \rho(\vec{u}') = \rho P(\vec{u}')$. The pair correlation function $g(\vec{r}, \vec{u}, \vec{u}')$ measures the deviation from this expectation at closer separations. Indeed, the central particle influences the positions and orientations of nearby particles because it interacts with them. The pair correlation function is, in essence,

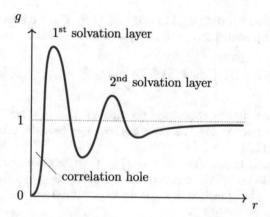

Fig. 3.4 The pair correlation g as a function of the distance r between the centres of mass of two particles. For short separations, $g \to 0$ due to volume exclusion, giving rise to the "correlation hole" (indicated). For large separations, $g \to 1$, while for intermediate separations, there are peaks and valleys associated with solvation layers (also indicated). Onsager theory for lyotropic nematics focuses on the effects of volume exclusion, so the correlation hole, while Maier–Saupe theory for thermotropics mainly deals with the effects of the attractive part of the interaction that contributes to the anisotropy of the solvation layer.

the ratio between the actual number density of particles near a test particle and the hypothetical value it would have had if the particles had not interacted.

At very short separations, we expect $g \to 0$ because two particles cannot occupy the same volume in space. This produces a "correlation hole" in $g(r)$ for positions where other particles are excluded due to the presence of reference particle at the origin of the co-ordinate system. For large separations, the influence of the central test particle becomes weak and $g \to 1$. At intermediate distances, there are peaks and valleys, associated with the so-called *solvation layers* of particles around the test particle. See Fig. 3.4. We naïvely expect the first (primary) peak to correspond to the minimum of the interaction potential: see Fig. 1.4. The reason is that g is a probability, and for classical particles this probability must be Boltzmann distributed. Hence, for low densities, g must be equal to $\exp(-U/k_B T)$. In reality, that is, at non-zero densities, the relevant potential is not the pair potential but the position-dependent, effective many-body interaction experienced by the test particle for a given orientation in three-dimensional space.[4]

For a uniform isotropic or nematic fluid, the molecular field any particle experiences from all other particles becomes

$$U_{\text{mol}}(\vec{u}) = \rho \int d\vec{u}' \, P(\vec{u}') \int d\vec{r} \, g(\vec{r}, \vec{u}, \vec{u}') U(\vec{r}, \vec{u}, \vec{u}'), \qquad (3.8)$$

---

[4] For hard particles, the value of the pair correlation function at contact is related to the bulk pressure [1].

which follows from inserting Eq. (3.7) into Eq. (3.6). This we insert in Eq. (3.5) to finally obtain the potential energy of our system of non-spherical particles,

$$U_{\text{pot}} = \frac{1}{2} N \rho \int d\vec{u} \int d\vec{u}' \, P(\vec{u}) P(\vec{u}') \int d\vec{r} \, g(\vec{r}, \vec{u}, \vec{u}') U(\vec{r}, \vec{u}, \vec{u}'). \qquad (3.9)$$

We can also write this in short-hand notation as $U_{\text{pot}} = \frac{1}{2} N \rho \langle \langle \int d\vec{r} \, g(\vec{r}, \vec{u}, \vec{u}') U(\vec{r}, \vec{u}, \vec{u}') \rangle \rangle'$, where the brackets represent averages over the orientations $\vec{u}$ and $\vec{u}'$. See also Eq. (2.3).

To calculate an excess free energy, F_{exc}, from this somewhat intimidating integral, we make use of the thermodynamic identity $U_{\text{pot}} = \partial \beta F_{\text{exc}} / \partial \beta$, where $\beta \equiv 1/k_B T$ [J^{-1}] is the reciprocal thermal energy.[5] It follows that

$$\beta F_{\text{exc}} = \frac{F_{\text{exc}}}{k_B T} = \int d\beta U_{\text{pot}}. \qquad (3.10)$$

This expresses the excess free energy of the system, F_{exc}, in terms of the orientational distribution function $P(\vec{u})$ and the pair correlation function $g(\vec{r}, \vec{u}, \vec{u}')$, both of which are unknown functions. This, of course, constitutes a not so minor inconvenience. One other inconvenience is that for simplicity we choose the integral to be undetermined and that we shall need to worry about constants of integration.

Our purpose will be to *calculate* the orientational distribution function from the overall free energy Eq. (3.1), given plausible educated guesses for either the *pair distribution function* or the *molecular field*.[6] One of these educated guesses focuses on the rôle of Van der Waals interactions and, in essence, accounts for the anisotropy in the first peak of the pair distribution function. See Fig. 3.4. This will turn out to be relevant to thermotropic nematics. The other approach ignores the impact of attractive interactions completely and presumes that the Pauli (volume) exclusion principle dominates the physics of liquid crystals. The volume exclusion gives rise to what we earlier called the "correlation hole" in the pair correlation function for particle configuration where their "hard" cores overlap, also shown in Fig. 3.4. This turns out to be accurate in the context of lyotropic liquid crystals of very long, rod-like particles.

The first approach gives rise to the Maier–Saupe theory of thermotropic nematics, and the second to the Onsager theory. In spite of the obvious differences, these two ways of dealing with the complexity of the structure of fluids are intimately connected and lead to equations for the orientational distribution function that, in

---

[5] Note that the internal energy E of a system obeys $\partial \beta F / \partial \beta$, with F the Helmholtz free energy and $\beta = 1/k_B T$ the usual reciprocal thermal energy $k_B T$. See also exercise 1 below. Note that E includes the kinetic and potential energy. We denote the latter with the symbol U in this book.

[6] Liquid-state theory is awash with educated guesses of pair correlation functions and/or functions derived from these, and known as *direct correlation functions*. These were introduced by Leonard Ornstein and Frits Zernike in 1914 in order to deal with the long-range nature of the pair correlation function near a critical point [1].

the end, look surprisingly similar. Let us first get to grips with Onsager theory in the next chapter and postpone an exposition of the Maier–Saupe theory to the chapter after the next.

Further Reading A clear and succinct presentation of the framework of classical thermodynamics can be found in the textbook of David Chandler [2]. This book also gives a didactic exposition of how to derive the Gibbs entropy formula and how a quantum-statistical description naturally leads to the (quasi) classical free energy of an ideal gas. For an intuitive view on half-open ensembles, relevant in the context of solutions and dispersions, the reader may refer to the textbook of Jean-Louis Barrat and Jean-Pierre Hansen on simple and complex fluids [3]. Incidentally, both textbooks provide a good introduction to correlation functions.

Exercises

1. **Ideal gases of point particles revisited.** The free energy of an ideal gas, F_{id}, is given by Eq. (3.2), where $\omega = \Lambda^3$ with $\Lambda = \sqrt{h^2/2\pi m k_B T}$ the thermal wavelength. Here, h [J s] denotes Planck's constant, m [kg] the mass of the particle, k_B [J K^{-1}] Boltzmann's constant, and T [K] the absolute temperature.

   (a) Verify that the ideal free energy, F_{id}, indeed produces the ideal gas law for the pressure p.

   (b) Derive from the free energy an expression for the chemical potential μ of ideal point particles.

   (c) Verify that the ideal gas pressure and ideal gas chemical potential obey the well-known Maxwell relation $\partial\mu/\partial V|_{N,T} = -\partial p/\partial N|_{V,T}$ from classical thermodynamics.

   (d) Show explicitly that the thermodynamic relation $\partial(F/T)/\partial(1/T) = E$ holds, if E represents the internal energy of the system at hand. Hint: Apply the chain rule for differentiation and insert the identities $S = -\partial F/\partial T$ and $F = E - TS$.

   (e) Use the relation $\partial(F/T)/\partial(1/T) = E$ to calculate the internal energy of a gas of ideal point particles using the ideal gas expression for the free energy. What kind of energy does this represent?

   (f) Calculate the heat capacity C_V at constant volume V of an ideal gas of point articles, and verify that it is an *invariant* of the temperature, i.e., does not depend on it.

2. **Ideal polar particles in an external field.** The orientational free energy of a collection of N non-spherical particles is given by Eq. (3.4), at least if we pretend that their orientations are not correlated. If the particles are polar, they can be aligned in an electric field. This means that their polar angle, relative to the field direction, is biased towards small angles. Let us for simplicity presume the orientational distribution function of the particles in the external field to obey a

step function, that is, $P(\vec{u}) = P(\theta, \phi) = [2\pi(1 - \cos\theta_0)]^{-1}$ for polar angles $0 \le \theta \le \theta_0$ and $P(\theta, \phi) = 0$ for angles $\theta_0 < \theta \le \pi$. There is no preference for any azimuthal angle $\phi \in [0, 2\pi]$. The cut-off angle $\theta_0 \in [0, \pi]$ is smaller the larger the field strength.

(a) Verify that the orientational distribution function is properly normalised:
$\int d\vec{u} P(\vec{u}) = \int_0^\pi d\theta \sin\theta \int_0^{2\pi} d\phi P(\theta, \phi) = \int_{-1}^{+1} d\cos\theta \int_0^{2\pi} d\phi P(\theta, \phi) = 1$.

(b) Calculate the expectation value $\langle \cos\theta \rangle$ of $\cos\theta$ as a function of θ_0. Compare the limits $\theta_0 = \pi$ for zero field, and $\theta_0 \ll 1$ for very strong fields, and argue why $\langle \cos\theta \rangle$ would be a good order parameter for the particles.

(c) Show that the variance $\sigma^2_{\cos\theta}$ of $\cos\theta$, defined as $\langle (\cos\theta - \langle\cos\theta\rangle)^2 \rangle$, can also be written as $\langle \cos^2\theta \rangle - \langle \cos\theta \rangle^2$.

(d) Evaluate the variance of $\cos\theta$, $\sigma^2_{\cos\theta}$, in the limits $\theta_0 = \pi$ and $\theta_0 \to 0$. What can be said about the uncertainty of the variable $\cos\theta$ in these two limits?

(e) Calculate the orientational entropy of the particles in the limits $\theta_0 = \pi$ and $\theta_0 \ll 1$. What can be said about the difference in entropy between these two cases?

3. **Lennard–Jonesium.** Consider a gas of interacting spherical particles. At low densities, the pair correlation function of such a gas can be approximated by a Boltzmann factor, $g(\vec{r}) = g(r) \approx \exp(-U(r)/k_B T)$. Here, U [J] denotes the interaction potential that depends only on the distance $r \equiv |\vec{r}|$ [m] between the particles, if we place one particle in the origin of the Cartesian co-ordinate system and the other at position $\vec{r}$. We presume the particles to interact via a Lennard–Jones potential, given in Eq. (1.5).

(a) Sketch of the Lennard–Jones potential. Calculate for that purpose (1) for what inter-particle distance the interaction potential is zero, (2) at what distance the potential has a minimum, and (3) what the value is of the potential at this distance.

(b) Indicate for what distances the potential must be repulsive and for what distances it is attractive.

(c) Plot the pair correlation function for values of the scaled energetic parameter $\epsilon/k_B T$ of 0.1, 1, and 2. Scale the distance between the particles to the parameter σ.

(d) What can be concluded about the rôle of temperature in the structuring fluids?

(e) Give a physical interpretation of the parameter σ.

(f) Argue why in the high-temperature limit $T \gg \epsilon/k_B T$ we may approximate the Lennard–Jones potential by a hard-core potential for which $U(r) \to \infty$ if $r \le \sigma$ and $U(r) = 0$ if $r > \sigma$. See also Chap. 2. What does the pair correlation function look like for this potential, if it approximately obeys the simple Boltzmann distribution?

References

1. J.-P. Hansen and I.R. McDonald, *Theory of simple liquids - with applications to soft matter*, 4th edition (AP, Amsterdam, 2013).
2. D. Chandler, *Introduction to modern statistical mechanics* (OUP, Oxford, 1987).
3. J.-L. Barrat and J.-P. Hansen, *Basic concepts for simple and complex fluids* (CUP, Cambridge, 2003).

Chapter 4
Onsager Theory

Abstract The free energy functional introduced in Chap. 3 can be made manageable, by inserting an educated guess for the pair correlation function. It produces an excess free energy identical to the one derived by Lars Onsager for hard rods, that is, for infinitely rigid, rod-like particles that interact via a harshly repulsive excluded volume interaction. Predictions of the Onsager theory are discussed.

As already alluded to, lyotropic nematics are dispersions of colloidal particles or macromolecules dispersed in a fluid medium. Lars Onsager announced in a brief abstract in 1942 that he could explain experimental observations of the co-existence of isotropic and nematic dispersion of rod-like and of plate-like colloidal particles, solely by accounting for their "co-volumes," that is, their mutually excluded volumes, in a thermodynamic description [1]. In 1949, he published the full paper, now a classic in the field, showing that entropy alone can drive phase transitions [2]. This constituted a major paradigm shift in our understanding of phase transitions.

Since, it has become clear that entropy-driven transitions abound in lyotropic systems and explain not only the existence of nematic liquid crystals, but also cholesteric, smectic, and columnar liquid crystals. An interesting consequence is that interfacial tensions between such phases must be entropy-dominated too. This goes against the grain of how we usually think about what brings about an interfacial tension, which is typically explained in terms of energy rather than entropy. All of this is at least as counter-intuitive as suggesting that colloidal crystals have a *larger*, not *lower*, entropy than the corresponding fluid colloid phase of the same colloid density. (Recall exercise 3 of Chap. 2.) Indeed, we would expect any crystal phase to be lower in entropy compared to the corresponding fluid phase on account of its lower symmetry.

The account of the Onsager theory that we give in this chapter is rather different from the original derivation, which is based on a virial expansion of the free energy. A formal derivation of the virial expansion of the free energy is somewhat technical and perhaps not all that intuitive to a non-expert in statistical mechanics. Onsager's version, which involves only the correction to first order in the density of the free energy of non-interacting particles, can be derived more intuitively, but the procedure is quite subtle and involves an assumption that, essentially, collisions between pairs of particle are uncorrelated. Our presentation is slightly more pedestrian, and, more importantly, it allows us to naturally connect Onsager theory and Maier–Saupe theory.

To start, let us presume that the particles are infinitely rigid cylinders of length L and width D, illustrated in Fig. 4.1. Pairs of cylinders interact via a steeply repulsive potential such as that shown in Fig. 4.2. The pair correlation function $g(\vec{r}, \vec{u}, \vec{u}')$ depends on the centre-to-centre distance, $\vec{r}$, as well as the orientations of the two particles, $\vec{u}$ and $\vec{u}'$. Referring to our discussion in the preceding chapter, we expect that $g \approx \exp(-\beta U)$ to lowest order in the density. If U is a steeply repulsive potential like the one shown in Fig. 4.2 (left), that is, if U is zero for non-overlapping

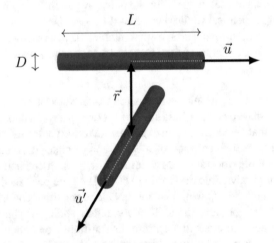

Fig. 4.1 Cylindrical particles of length L and width D, and orientations $\vec{u}$ and $\vec{u}'$. We presume the interaction potential to depend steeply on the shortest separation, $r \equiv |\vec{r}|$ between the centre lines of the particles.

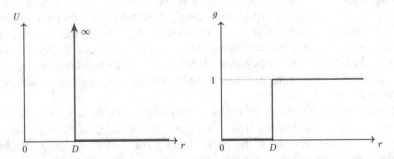

Fig. 4.2 Hard-core potential U (left) and the corresponding pair correlation function g (right) as a function of the distance between the centre lines r for an ideal gas of "hard" rods, i.e., cylinders of width D. See also Fig. 4.1.

particles and infinity for overlapping ones, then the pair correlation becomes a step function, schematically represented in the same Fig. 4.2 (right).

So, the only feature of the pair correlation function that we keep is what in the preceding chapter we called the *correlation hole*. Superficially, this seems a crude approximation but in fact works very well at the low densities where nematic phases arise in dispersions of long rod-like particles. Of course, if the colloidal particles are charged, then the Coulombic interaction potential is much softer and the correlation hole not as sharp. It turns out, however, that to approximately account for electrostatic interactions, we only need to "renormalise" (adjust) the radius, that is, make it slightly larger. By how much larger, depends on the charge density on the surface of the particle and the concentration of added salt. This was in fact already suggested by Onsager in his ground-breaking 1949 paper. Strictly speaking, this is not quite the whole story, as it misses an important twisting effect caused by the electrostatic interaction. For our purposes, it suffices to know that a hard-core potential is a good representative of actual repulsive interactions between colloidal particles.

If we take $g = \exp(-\beta U)$ as our estimate for the pair correlation function, then the potential energy of our collection of rod-like particles becomes

$$U_{\text{pot}} \approx \frac{1}{2}N\rho \int d\vec{u}\, P(\vec{u}) \int d\vec{u}'\, P(\vec{u}') \int d\vec{r}\, U(\vec{r}, \vec{u}, \vec{u}') \exp\left(-\beta U(\vec{r}, \vec{u}, \vec{u}')\right).$$

$$(4.1)$$

Notice that the interaction potential U has to drop sufficiently fast to zero with inter-particle distance for the spatial integral not to diverge in the thermodynamic limit $V \to \infty$. We do not actually insert a hard-core model potential $U \to \infty$ for overlapping configurations and $U = 0$ for non-overlapping configurations of particles in Eq. (4.1), at least not yet. Indeed, on the face of it, Eq. (4.1) seems to produce zero if we did that! This, in fact, is what we should expect, for overlapping configurations would have zero Boltzmann probability and the expectation value of the interaction energy must indeed be zero.

However, integrating in Eq. (4.1) over β gives us, according to Eq. (3.10), our excess free energy, which includes contributions from the entropy loss, associated with the existence of the correlation hole. This excess free energy is well-behaved, even for hard-core interactions, and reads

$$\beta F_{\text{exc}} = \frac{1}{2}N\rho \int d\vec{u} \int d\vec{u}'\, P(\vec{u}) P(\vec{u}') \int d\vec{r} \left[1 - \exp\left(-\beta U(\vec{r}, \vec{u}, \vec{u}')\right)\right], \quad (4.2)$$

if we use the integration constant in the quantity β to make certain that for $U = 0$ we retrieve the expected $U_{\text{pot}} = 0$.[1] Equation (4.2) can be rewritten as

$$\beta F_{\text{exc}} = N\rho B, \tag{4.3}$$

where B [m^3] is the second virial co-efficient, and we recall that ρ is the number density of the particles. Hence, $B\rho$ is dimensionless, as it should. For cylindrical particles, it is defined in terms of orientational averages $\langle \cdots \rangle \equiv \int d\vec{u}\, P(\vec{u})(\cdots)$, with a similar definition for the primed variable, as

$$B \equiv \langle\langle B(\vec{u}, \vec{u}')\rangle\rangle'. \tag{4.4}$$

Here, the second virial co-efficient of two cylindrical particles with fixed orientations $\vec{u}$ and $\vec{u}'$ is defined as

$$B(\vec{u}, \vec{u}') \equiv \frac{1}{2} \int d\vec{r} \left[1 - \exp\left(-\beta U(\vec{r}, \vec{u}, \vec{u}')\right)\right]. \tag{4.5}$$

From this definition of $B(\vec{u}, \vec{u}')$, and realising that the integrand (the function that is to be integrated) has a value of unity for overlapping configuration of pairs of hard cylinder and zero for all other cases, we find, following Lars Onsager, or using a cunningly chosen co-ordinate system[2] due to Joseph Straley [4],

$$B(\vec{u}, \vec{u}') = \frac{1}{2} \times L^2 |\sin \gamma(\vec{u}, \vec{u}')| \times 2D + \mathcal{O}(LD^2), \tag{4.6}$$

with $\gamma(\vec{u}, \vec{u}') = \arccos(\vec{u} \cdot \vec{u}')$ the angle between the main body axis vectors of two cylinders, and $\cos \gamma = \cos\theta \cos\theta' + \sin\theta \sin\theta' \cos(\phi - \phi')$. Here, θ and θ' denote usual the polar angles, and ϕ and ϕ' the azimuthal angles, of the main body axis vectors of the two cylindrical particles. In Eq. (4.6) we ignore terms of order $LD^2 \ll L^2D$ indicated by the symbol $\mathcal{O}(\cdots)$. These terms represent interaction of an end of one particle with a cylindrical centre of another particle, and between the ends of two particles.

Equation (4.6) is actually quite intuitive, if we keep in mind Fig. 4.1. It suggests that the volume excluded by two rods that are skewed at some angle γ must approximately be equal to the area of a parallelogram with equal sides of length L and opening angle γ, times two times the width D of the rods. Hence, $B(\vec{u}, \vec{u}')$ is

---

[1] The ruse that we apply here is inspired by that of Ben Widom in a very similar context [3]. If uncomfortable with this procedure, use a constant finite potential with $U = \epsilon > 0$ for overlapping and $U = 0$ for not overlapping configurations and take the limit $\epsilon \to \infty$ at a later stage.

[2] The relative position of the particles can be described by an oblique co-ordinate system spanned by the main body axis vectors, $\vec{r} \to \xi\vec{u} + \eta\vec{u}' + \zeta\vec{u} \times \vec{u}'/|\vec{u} \times \vec{u}'|$, giving a volume element $d\vec{r} \to d\xi d\eta d\zeta |\vec{u} \times \vec{u}'| = d\xi d\eta d\zeta |\sin \gamma(\vec{u}, \vec{u}')|$. Here, one particle is pinned at the origin, and the other at $\vec{r}$. The integrand is unity for overlapping rods only and zero otherwise, producing non-zero contributions for values of the co-ordinates $|\xi|, |\eta| < L/2$ and $|\zeta| < D$.

half the volume of the correlation hole that is equal to that of a parallelepiped with base area $L^2|\sin\gamma(\vec{u},\vec{u}')|$ and height $2D$. Aligning rods reduces the second virial co-efficient B, and we conclude that the excess free energy F_{exc} must in that case decrease. Consequently, the free volume accessible to all particles in the volume increases if the particles align along some common axis. Obviously, this happens at the cost of orientational entropy and leads to an increase of the orientational free energy $F_{or} = Nk_BT\langle\ln P\rangle$ (see Chap. 4).

To complete the Onsager theory, we add all ingredients to our free energy functional $F = F_{id} + F_{or} + F_{exc}$, and obtain

$$\frac{\beta F[P]}{N} = \ln\rho\omega - 1 + \int d\vec{u}\, P(\vec{u})\ln P(\vec{u}) + \rho\int d\vec{u}\int d\vec{u}'\, P(\vec{u})P(\vec{u}')B(\vec{u},\vec{u}'),$$

(4.7)

which may also be written in short-hand notation as

$$\frac{\beta F[P]}{N} = \ln\rho\omega - 1 + \langle\ln P\rangle + \rho\langle\langle B(\vec{u},\vec{u}')\rangle\rangle' = \ln\rho\omega - 1 + \langle\ln P\rangle + \rho B, \quad (4.8)$$

where we recall that ω is volume scale that depends on the temperature and the chemical potential of the solvent (see Chap. 3). Because this free energy only accounts for the contribution of the second virial co-efficient, and higher order virials involving three and higher order contacts ("collisions") are ignored, the theory is an example of a second virial theory. Interestingly, there are strong indications that the second virial approximation becomes exact in the limit of infinite aspect ratio $L/D \to \infty$ [5, 6]. This makes Onsager theory one of the few exact theories in condensed matter physics. In practice, the theory is for all intents and purposes quantitatively correct for aspect ratios in excess of a few hundred.

In thermodynamic equilibrium, the most probable distribution P minimises the free energy F. This is a consequence of the second law of thermodynamics but in fact also follows from the ensemble theory of statistical mechanics. So, we have to minimise the free energy but in the process have to make certain that we do not violate the condition of the normalisation of probability distributions, $\int d\vec{u}\, P(\vec{u}) \equiv 1$. This we do using the *method of Lagrange multipliers* that we explain next, taking for simplicity as example some function $f(x, y)$ of the pair of variables x and y.

To find a maximum or minimum of the function $f(x, y)$ and demand that x and y obey the equality $g(x, y) = 0$, it is not sufficient to demand that f is a stationary function of these variables x and y. For f to be stationary, we must insist that $df = f_x dx + f_y dy = 0$, where $f_x \equiv \partial f/\partial x$ and $f_y \equiv \partial f/\partial y$ are partial derivatives. For arbitrary variations dx and dy, both derivatives have to be equal to zero at a stationary point. To find a stationary point under the constraint that $g(x, y) = 0$, and noting that because of that $dg = g_x dx + g_y dy = 0$, we need to demand that $f - \lambda g$ is stationary and hence that $d(f - \lambda g) = 0$, where λ is known as a Lagrange multiplier. We fix the value of multiplier *post hoc*, so afterward, by inserting the solution in the function g and demanding that $g = 0$. The methodology generalises straightforwardly to multivariate functions, that is, functions of many variables.

In our case, the minimisation is a so-called *functional minimisation* and indicated by curly deltas. Demanding the free energy to be an extremum translates to the equality

$$\frac{\delta}{\delta P}\left[\frac{\beta F}{N} - \lambda\left(\int d\vec{u}\, P(\vec{u}) - 1\right)\right] = 0, \tag{4.9}$$

where λ is a Lagrange multiplier introduced to ensure proper normalisation of the distribution function. A pedestrian recipe for obtaining the functional derivative is as follows. First, insert $P(\vec{u}) + \delta P(\vec{u})$ into the expression for the free energy, Eq. (4.7), where $P(\vec{u})$ is the as yet unknown most probable distribution function and $\delta P(\vec{u})$ a perturbation away from it. Second, Taylor expands the integrands in the free energy functional to linear order in $\delta P(\vec{u})$. See equation 10 of exercise 3 of Chap. 2 for a reminder of how to Taylor expand a function. Finally, rearrange the expressions to produce an expression of the form $F[P + \delta P] = F[P] + \int d\vec{u}\delta P(\vec{u})\,[\delta F/\delta P]$, defining the functional derivative $\delta F/\delta P$ as all terms that end up between the square brackets in this expression.

In our case, the first integral produces

$$\int d\vec{u}\left[P(\vec{u}) + \delta P(\vec{u})\right]\ln\left[P(\vec{u}) + \delta P(\vec{u})\right] = \int d\vec{u}\, P(\vec{u})\ln P(\vec{u})$$

$$+ \int d\vec{u}\delta P(\vec{u})\left[\ln P(\vec{u}) + 1\right]$$

$$+ \cdots \tag{4.10}$$

up to first order in the perturbation δP. The second integral becomes

$$\rho\int d\vec{u}\int d\vec{u}'\left[P(\vec{u}) + \delta P(\vec{u})\right]B(\vec{u}, \vec{u}')\left[P(\vec{u}') + \delta P(\vec{u}')\right]$$

$$= \rho\int d\vec{u}\int d\vec{u}'\left[P(\vec{u})B(\vec{u}, \vec{u}')P(\vec{u}') + 2\delta P(\vec{u})B(\vec{u}, \vec{u}')P(\vec{u}')\right] + \cdots, \tag{4.11}$$

again to first order in δP. The factor of two arises because we can make use of the co-ordinate transformation $\vec{u} \rightarrow \vec{u}'$ and $\vec{u}' \rightarrow \vec{u}$, the symmetry relation $B(\vec{u}, \vec{u}') = B(\vec{u}', \vec{u})$, and the fact that the integration order is immaterial.

The first terms in Eqs. (4.10) and (4.11), together with the ideal gas term, represent $\beta F[P]/N$. The remainder can be written in the form $N\int d\vec{u}\delta P(\vec{u})$ $[\delta(\beta F/N)/\delta P]$, producing from Eq. (4.9) what in the field is known as the *Onsager equation*,

$$\ln P(\vec{u}) + 1 + 2\rho\int d\vec{u}'\, P(\vec{u}')B(\vec{u}, \vec{u}') - \lambda = 0, \tag{4.12}$$

using the fact that $\delta \int d\vec{u} P(\vec{u})/\delta P = 1$ [7]. This equation needs to be solved in order to get the equilibrium distribution $P(\vec{u})$. The value of λ can be found by normalising P. So far, no-one has succeeded in solving this non-linear integral equation analytically, except for the trivial solution $P = 1/4\pi$, which is the isotropic solution. We have to take recourse to numerical methods to find non-isotropic solutions, albeit that scaling laws in the limit of strongly ordered nematics have been derived [8, 9]. The reader is referred to the didactic review of the Onsager equation by René van Roij for how to solve the Onsager theory numerically, and how to use the numerical solution to obtain information on the thermodynamics of the dispersion of hard rods.

Even without numerically solving the Onsager equation, we are able to deduce that P obeys, as expected, a Boltzmann distribution,

$$P(\vec{u}) \propto \exp\left[-2\rho\langle B(\gamma)\rangle'\right] \equiv \exp\left[-\beta U_{\mathrm{mol}}(\vec{u})\right] \tag{4.13}$$

where $\beta U_{\mathrm{mol}}(\vec{u})$ is the earlier-introduced molecular field. This molecular field apparently depends on the distribution function itself. Hence, equations of this type are known as *self-consistent field equations*, and any theory that produces such an equation is a *self-consistent field theory*. Self-consistent field theories are mean-field theories.

Another useful observation that we can make from the Onsager equation is that the expression $\rho\langle B(\gamma)\rangle' \sim L^2 D\rho\langle|\sin\gamma|\rangle'$, valid to leading order in $L/D \gg 1$, suggests a natural dimensionless concentration scale $c \equiv \pi L^2 D\rho/4$. This is because the isotropic solution to Eq. (4.13) reads $P = 1/4\pi$, and that in that case $\langle|\sin\gamma|\rangle' = \pi/4$. It is instructive to rewrite this concentration $c = \phi L/D$ in terms of the volume fraction ϕ of the particles and their aspect ratio L/D.[3] It shows that c can be larger than unity, even if the volume fraction of the particles is small, with $\phi \ll 1$, provided they are sufficiently slender and $L/D \gg 1$. In that case, we can have what we perceive as a dilute solution but where the particles do sense each other's presence, nonetheless. It is the consequence of the circumstance that the volume of a rod-like particle is much smaller than its excluded volume, and more so the larger its aspect ratio.

If we perform a formal bifurcation analysis of the Onsager equation, we find that other solutions to the Onsager equation branch off from the isotropic solution $P = 1/4\pi$ at $c = c_{\mathrm{I}}^{\mathrm{spinodal}} = 4$, which turns out to be the isotropic spinodal [10]. For a pedestrian version of such an analysis, we refer to exercise 3 below where we also demonstrate that the branch point is the spinodal of the isotropic phase.

All of this implies that we must expect the dimensionless concentrations of the isotropic and co-existing nematic phases to be close to the spinodal value of $c = 4$. As a consequence, the transition to the nematic phase occurs around a volume fraction of $4D/L \to 0$ if $L/D \to \infty$, explaining why nematic phases can

---

[3] The precise correspondence depends on whether the rods are cylinders or spherocylinders, i.e., capped cylinders. In the limit $L/D \gg 1$, the distinction becomes insignificant.

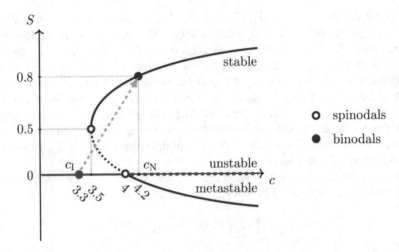

Fig. 4.3 Schematic of the phase diagram of dispersions hard rods in terms of the scalar nematic order parameter S as a function of the dimensionless concentration c. Indicated are the spinodals as well as the binodals: the limits of thermodynamic stability of the isotropic and nematic phases, and conditions of thermodynamic equilibrium.

present themselves at very low particle concentrations. A schematic of the numerical solution to the Onsager equation is given in Fig. 4.3. Plotted is the order parameter $S = \langle P_2(\cos\theta)\rangle$ as a function of the dimensionless concentration c. For $c < 3.5$ there is only one solution to the Onsager equation. For $c > 3.5$ there are three. Two of these are free energy minima, and one is a free energy maximum.[4] The free energy maxima for $c < 4$ and $c > 4$ are indicated by the dashed lines. Stable branches end in the limits of thermodynamic stability known as the isotropic and nematic spinodals that are also indicated in the figure.

The isotropic spinodal is located at $c = 4$, and the nematic spinodal at $c = 3.5$. At these points, the isotropic and nematic solutions to the Onsager equation become unstable with respect to small perturbations in the degree of orientational order. Actually, the nematic branch becomes *mechanically* unstable for concentrations below 3.7, for which the osmotic compressibility becomes negative [10, 12]. The branch with negative order parameter represents a local, that is, *metastable* minimum. At equal concentration, the free energy of the ordered phase with positive order parameter is lower than that with negative order parameter.

To pinpoint the binodal, that is, find out what the concentrations of particle in the co-existing isotropic (I) and nematic (N) phases are, we need to make use of the second law of thermodynamics. According to the second law, any two phases in thermodynamic equilibrium must be in chemical, mechanical, and thermal equilibrium. This means equal chemical potentials, $\mu_{\mathrm{I}} = \mu_{\mathrm{N}}$ with $\mu =$

---

[4] To ascertain that solutions to the Onsager equation are minima, we have to investigate if the second functional derivative of the free energy has a positive definite norm [11].

$\partial F/\partial N|_{V,T}$, equal osmotic pressures, $\Pi_I = \Pi_N$ with $\Pi = -\partial F/\partial V|_{N,T}$, and equal temperatures, $T_I = T_N$. Since the theory describes the balancing of two types of entropy and has no energetic component, it is athermal.[5] Consequently, there is no meaningful temperature in this model, and we only need to equate chemical potentials and osmotic pressures in both phases. We do not reproduce the expressions for these thermodynamic quantities here and refer to the expressions provided in exercise 2 below.

A numerical evaluation of the governing set of equations gives for the concentration in the isotropic phase $c_I \simeq 3.3$ and a corresponding order parameter of $S = 0$, while for the concentration in the nematic phase we obtain $c_N \simeq 4.2$ with order parameter $S \simeq 0.8$. Hence, not only are the particles aligned in the nematic phase, the concentration of particles is also about 30% larger in the nematic than that in the co-existing isotropic phase. These values agree favourably with extrapolations from results from computer simulations and explain why under conditions of co-existing isotropic and nematic phases in a properly equilibrated sample, the bottom phase in a crucible tends to be the nematic phase [13]. See Fig. 2.9.

The question arises how precisely to read the phase diagram of Fig. 4.3. At low concentrations, that is, for $c \leq 3.3$, the isotropic phase is the stable phase. On the other hand, for $c \geq 4.2$, the nematic phase is the thermodynamically stable state of the dispersion. The nematic order parameter $S \geq 0.8$ increases with increasing concentration and approaches the value of unity only at very high densities.[6] (In reality, the nematic phase transitions to the smectic A phase before that this happens.) Preparing the dispersion at an overall concentration of particles in the region $3.3 < c < 4.2$ produces two co-existing phases: one isotropic phase with concentration $c = c_I = 3.3$ and order parameter $S = 0$, and one nematic phase with concentration $c = c_N = 4.2$ and order parameter $S = 0.8$. The fractional volume V_N/V of the nematic phase is fixed by the overall concentration: $V_N/V = (c - c_I)/(c_N - c_I)$.[7] It increases from $V_N/V = 0$ if $c \leq c_I$ with increasing concentration to $V_N/V = 1$ if $c \geq c_N$. See Fig. 4.4. As already alluded to, the branch with negative order parameter is metastable because the free energy is always larger than that of the corresponding nematic with positive order parameter (at equal density).

Finally, we need to realise that the phase diagram of Fig. 4.3 is *universal*, that is, valid for any type of cylindrical colloidal particle of which the interaction can be represented by a hard-core repulsive potential, irrespective of its length L and width D. So, it actually represents a *law of corresponding states*, similar to the one we discussed in Chap. 2 in the context of the condensation of gases. In fact, the shape of the bifurcation diagram near the nematic spinodal is very similar to

---

[5] In a way, we are in some effective high-temperature limit, where energy is irrelevant and entropy rules.

[6] High degrees of order can be achieved also in the presence of strong magnetic fields [14, 15].

[7] This is true only for monodisperse particles. For polydisperse particles, the situation is more complicated [16].

Fig. 4.4 Schematic representation of how isotropic and nematic phases present themselves in a lyotropic system. Left: for particle concentrations below where the nematic becomes stable, we observe only one phase, which is the isotropic phase. Right: for concentrations larger than the minimum concentration needed to stabilise the ordered phase, again only one phase (the nematic phase) is seen. Middle: for intermediate concentrations, we have co-existence of an isotropic phase and a nematic phase. Their relative volumes depend on the overall concentration of the particles in the dispersion.

the binodal curve of the gas–liquid transition near the critical point! We recall that within Onsager theory, the nematic spinodal is located at the dimensionless concentration $c = c_N^{\mathrm{spinodal}} = 3.5$, with an associated scalar nematic order parameter $S = S_N^{\mathrm{spinodal}} = 0.4$.[8] Indeed, near the nematic spinodal, we have $|S - S_N^{\mathrm{spinodal}}| \propto \sqrt{c - c_N^{\mathrm{spinodal}}}$. This is to be compared with Eq. (1.3) of the binodal of the co-existing gaseous and liquid states of simple compounds, noting that the appropriate order parameter is in that case the density. In the present case, the mean-field critical exponent of one-half is thought to be exact. Sufficiently elongated hard rods are one of the few known examples in condensed matter physics that obey mean-field statistics in three spatial dimensions! This means it does *not* belong to the universality class of gases and liquids, unlike, for instance, ferromagnets. Ferromagnets are solids such as iron and nickel that exhibit a transition between a disordered (paramagnetic) state and an ordered (ferromagnetic) state at a critical temperature called the Curie temperature.

This now brings us to the Maier–Saupe theory for thermotropic nematics, which, as we shall see, has a structure very similar to that of the Onsager theory, and also produces a law of corresponding states very similar to that of the Onsager theory. However, actual thermotropics turn out *not* to behave mean-field-like in three spatial dimensions, and this renders the theory qualitative rather than quantitative.

Further Reading For an in-depth yet concise theoretical review of Onsager theory and its extensions, the reader is referred to that of Theo Odijk [5]. For a more accessible yet very complete account of the theory of lyotropic liquid crystals and

---

[8] One could perhaps also call the point of mechanical instability for $c = 3.7$ where the compressibility diverges [10] the nematic spinodal, even though the free energy remains locally stable to perturbation in the orientational distribution function down to $c = 3.5$.

comparison with experiments, the reader may consult the review paper of Gert-Jan Vroege and Henk Lekkerkerker [6]. An accessible and slightly more recent review paper that deals more extensively with improvements upon Onsager theory for finite aspect ratios is that of Dave Williamson [17]. The textbook of Remco Tuinier and Henk Lekkerkerker gives a very intuitive account of an extension of Onsager's theory for hard rods of finite aspect ratio, based on the so-called y-expansion of Boris Barboy and Bill Gelbart [18]. A didactic treatment of density functional theory in the context of a host of soft matter-related problems is given the textbook of Jean-Louis Barrat and Jean-Pierre Hansen [19].

Exercises

1. **Dilute dispersion of dipolar rod-like particles in an external field.** Let us consider a dilute solution of rod-like particles. The particles carry a permanent dipole moment $\vec{p} = p\vec{u}$ along the main body axis vector $\vec{u}$. Here, $p = |\vec{p}|$ is the magnitude of the dipole moment, and we recall that $\vec{u}$ is a unit vector. If we switch on an electric field $\vec{E} \equiv E\vec{n}$ of magnitude $E = |\vec{E}|$ that points along the unit vector $\vec{n}$ (the "director"), then the electrostatic energy $U_{\text{ext}}(\vec{u})$ that a particle experiences is equal to $U_{\text{ext}}(\vec{u}) = -\vec{p} \cdot \vec{E} = -pE(\vec{u} \cdot \vec{n})$. If sufficiently dilute, we can ignore the contribution from the interactions between the particles. The overall free energy of the collection of N particles in the volume V of the solution is then given by the sum of the ideal free energy, the orientational free energy, and a contribution from the external field, $F = F_{\text{id}} + F_{\text{or}} + F_{\text{ext}}$, where the first two terms are given by Eqs. (3.2) and (3.4). The contribution from the external field amounts to $F_{\text{ext}} = N \int d\vec{u} P(\vec{u}) U_{\text{ext}}(\vec{u}) = N\langle U \rangle$, where $P(\vec{u})$ is the as yet unknown orientational distribution function.

   (a) Show by functionally minimising the overall free energy with respect to the orientational distribution function that the orientational distribution function obeys the familiar Boltzmann distribution

$$P(\vec{u}) = \frac{\exp\left[-\beta U_{\text{ext}}(\vec{u})\right]}{\int d\vec{u} \exp\left[-\beta U_{\text{ext}}(\vec{u})\right]}, \tag{4.14}$$

   under conditions of thermal equilibrium, with $\beta \equiv 1/k_B T$ the reciprocal thermal energy.

   (b) Verify that the normalisation of the orientational distribution function, also known as the single-particle *partition function* $Z \equiv \int d\vec{u} \exp\left(-\beta U_{\text{ext}}(\vec{u})\right)$,[9]

---

[9] Strictly speaking, a *configurational integral*, as it does not contain the integration over the momenta of the particles.

can be written as $Z = 4\pi K^{-1} \sinh K$ with $K \equiv \beta p E$, with the hyperbolic sine defined as $\sinh K \equiv [\exp(K) - \exp(-K)]/2$.

(c) Let the nth moment of the orientation of the particles along the field be defined by the quantity $\langle x^n \rangle \equiv \langle \vec{u} \cdot \vec{n} \rangle \equiv \langle (\cos \theta)^n \rangle$, with θ the angle between the main body axis vector of a particle and the "director." Demonstrate that it can be calculated from Z according to the simple relation

$$\langle x^n \rangle = \frac{1}{Z} \frac{d^n Z}{d K^n}, \tag{4.15}$$

for integers $n > 0$, and that for $n = 1$, we get for the mean degree of alignment $\langle x \rangle = -K^{-1} + \cosh K / \sinh K$, with $\cosh K \equiv [\exp(K) + \exp(-K)]/2$ the hyperbolic cosine of K.

(d) Plot as a function of the dimensionless field strength K, for values in the range from 0 to 20: the mean degree of alignment $\langle x \rangle$, the orientational free energy $\beta F_{or}/N$, and the electrostatic free energy $\beta F_{ext}/N$. Explain your findings.

(e) Show that free energy of the collection of dipoles is connected to the single-particle partition function Z by $F = F_{or} + F_{ext} = -N k_B T \ln Z$, as to be expected from statistical mechanics of non-interacting particles [20]. Show also that the *variance* $\sigma_x^2 \equiv \langle (x - \langle x \rangle)^2 \rangle = \langle x^2 \rangle - \langle x \rangle^2$ can be calculated from the partition function according to

$$\sigma_x^2 = \frac{d^2 \ln Z}{d K^2} = \frac{d \langle x \rangle}{d K}. \tag{4.16}$$

(f) Plot the standard deviation σ_x as a function of the expectation value $\langle x \rangle$, using K as a dummy variable, and choose for values of 0 to 10. Discuss what you find.

2. **Thermodynamics of isotropic dispersions of hard rods.** We now consider a dispersion of hard cylindrical particles of length L and width D. The particles are much longer than they are wide, $L \gg D$, and hence, the second virial approximation holds up to the concentration where the nematic phase appears, and, in fact, considerably beyond that.

(a) Verify, using the free energy Eq. (4.8), that the osmotic pressure Π of the dispersion of hard rods obeys

$$\beta \Pi = \rho + B\rho^2 \tag{4.17}$$

within the second virial approximation, where, as before, $B \equiv \langle\langle B(\vec{u}, \vec{u}') \rangle\rangle'$.

(b) Show that the chemical potential of the dispersion of hard rods, μ, is given by the expression

$$\beta \mu = \ln \rho \omega + \langle \ln P \rangle + 2\rho B \tag{4.18}$$

at the same level of approximation.

(c) Do the approximate expressions for the osmotic pressure Π and chemical potential μ obey the *exact* Maxwell relation $\partial\mu/\partial V|_{N,T} = -\partial\Pi/\partial N|_{V,T}$? Note that we tacitly presume the chemical potential of the solvent to be constant.

(d) Draw for the isotropic phase, for which $\langle\langle|\sin\gamma|\rangle\rangle' = \pi/4$, the compressibility factor $Z \equiv \beta\Pi/\rho$ as a function of the volume fraction $\phi = \pi\rho L D^2/4$ of the rod-like particles that we presume very elongated, with $L \gg D$. Estimate for what volume fractions the van 't Hoff law, for which $Z = 1$, is violated. Compare this with the volume fraction for which you expect the nematic phase to appear.

3. **The spinodal of the isotropic phase of hard rods.** The Onsager equation, Eq. (4.12), is a non-linear integral equation. Inserting Eq. (4.6) for the second virial co-efficient $B(\vec{u}, \vec{u}')$, we can write it in slightly different form

$$\ln P(\vec{u}) = \lambda - \frac{8c}{\pi}\langle|\sin\gamma|\rangle', \tag{4.19}$$

where we absorbed the number -1 in the multiplier λ. As before, γ is the angle between the rods of orientations $\vec{u}$ and $\vec{u}$, $\langle|\sin\gamma|\rangle' = \int d\vec{u}' P(\vec{u}')|\sin\gamma|$ and $c = \pi L^2 D\rho/4$ the dimensionless concentration. We have seen that it has more than a single solution for sufficiently large concentrations.

(a) Argue why $P(\vec{u}) = P$, with P a constant of $\vec{u}$, is a solution to the Onsager equation. Note that in that case $\langle|\sin\gamma|\rangle' = P\pi^2$. Fix the value of λ by insisting that P should be normalised and equal to $(4\pi)^{-1}$.

It turns out that because of the cylindrical symmetry of the isotropic and nematic phases, we can make use of what is known as the *addition theorem* for spherical harmonics [10]. It allows us to write

$$\int_0^{2\pi} d\phi|\sin\gamma| = 2\pi\left(\frac{\pi}{4} - \frac{5\pi}{32}P_2(x)P_2(x') + \cdots\right), \tag{4.20}$$

where $P_2(x) = (3x^2 - 1)/2$ denotes the earlier-introduced second Legendre polynomial, and $x = \cos\theta$ with θ the polar angle relative to the director that we take as the z-axis in a Cartesian co-ordinate system [10]. A similar prescription holds for the primed variable. We ignore contributions from higher order polynomials indicated by the dots $\cdots$, as we shall be interested in finding solutions that branch off from the trivial, isotropic solution, implying weak nematic order.

This transforms Onsager's integral equation into a more manageable form,

$$\ln P(x)/2\pi = \lambda - 2c\left[1 - \frac{5}{8}P_2(x)\int_{-1}^{+1} dx' P(x')P_2(x')\right], \tag{4.21}$$

where $P(\vec{u}) = P(x)/2\pi$ on account of the cylindrical symmetry, $P(x)$ is now normalised as $\int_{-1}^{+1} dx\, P(x) = 1$, and we absorb a constant $\ln 2\pi$ in λ. Other solutions branch off from the isotropic solution at some critical value of the scaled density c. To find this critical value, we write

$$P(x) = \frac{1}{2} + \varepsilon P_2(x) \qquad (4.22)$$

with $\varepsilon \to 0$ some small positive or negative number. Note that this trial function is properly normalised for arbitrary value of ε, because $\int_{-1}^{+1} dx\, P_2(x) = 0$. Note also that the perturbation has the symmetry of the nematic phase and could be seen as a "nematic" perturbation.

(b) Verify that $\int_{-1}^{+1} dx\, P_2^2(x) = 2/5$, so that the order parameter for the given distribution function must be equal to $S = 2\varepsilon/5$. This implies that for $\varepsilon > 0$ the order parameter is positive, while for $\varepsilon < 0$ it is negative.

(c) Show that Eq. (4.22) is a solution of Eq. (4.21) for $c = 4$, provided $\varepsilon \ll 1$. Hint: multiply left and right by $P_2(x)$, and integrate and Taylor expand the logarithm, using Eq. (1.10). Note that since Eq. (4.22) is normalised, the isotropic solution fixes the value of λ.

(d) If we insert the perturbative ansatz Eq. (4.22) for the distribution function in expression for the free energy, Eq. (4.7), subtract the free energy of the unperturbed isotropic distribution, we obtain an expression of the following form:

$$\beta \Delta F(S) = \frac{5}{2} N \left(1 - \frac{1}{4} c \right) S^2 + \cdots \qquad (4.23)$$

at least to quadratic order in S. Discuss what this means for the thermodynamic stability of the isotropic phase for $c < 4$ and $c > 4$, and why $c = 4$ must be the spinodal of the isotropic phase.

References

1. L. Onsager, *Anisotropic solutions of colloids*, Phys. Rev. **62** (1942), 558.
2. L. Onsager, *The effects of shape on the interaction of colloidal particles*, Ann. N. Y. Acad. Sci. **51** (1949), 627.
3. B. Widom, *Statistical mechanics - a concise introduction for chemists* (CUP, Cambridge, 2002);
4. J. P. Straley, *Frank elastic constants of the hard-rod liquid crystal*, Phys. Rev. **8** (1973), 2181.
5. T. Odijk, *Theory of lyotropic polymer liquid crystals*, Macromolecules **19** (1986), 2313.
6. G. J. Vroege and H. N. W. Lekkerkerker, *Phase transitions in lyotropic colloidal and polymer liquid crystals*, Rep. Prog. Phys. **55** (1992), 1241.
7. J.-P. Hansen and I.R. McDonald, *Theory of simple liquids - with applications to soft matter*, 4th edition (AP, Amsterdam, 2013).

8. R. van Roij and B. Mulder, *High-density scaling solution to the Onsager model of lyotropic nematics*, Europhys. Lett. **34** (1996), 201.
9. A. Poniewierski, *Nematic to smectic A transition in the asymptotic limit of very long hard spherocylinders*, Phys. Rev. A **45** (1992), 5605.
10. R. F. Kayser, Jr. and H.J. Raveché, *Bifurcation in Onsager's model of the isotropic-nematic transition*, Phys. Rev. **17** (1978), 2067.
11. J. Stecki and A. Kloczkowski, *On the stability of the orientational distribution of molecules*, J. Phys. Colloq. **40** (1979), C3–360.
12. S. P. Finner, *Transient clusters and networks in isotropic and symmetry-broken dispersions of slender nanoparticles* (PhD thesis, Eindhoven University of Technology, 2020).
13. H. N. W. Lekkerkerker and Remco Tuinier, *Colloids and the depletion interaction* (Springer, Dordrecht 2011).
14. M. Kleman and O. D. Lavrentovich (Springer, New York, 2003).
15. K. R. Purdy and S. Fraden, *Isotropic-cholesteric phase transition of filamentous virus suspensions as a function of rod length and charge*, Phys. Rev. E **70** (2004), 061703.
16. A. Speranza and P. Sollich, *Simplified Onsager theory for isotropic-nematic phase equilibria of length polydisperse hard rods*, J. Chem. Phys. **117** (2002), 5421.
17. D. C. Williamson, *The isotropic-nematic phase transition: the Onsager theory revisited*, Physica A **220** (1995), 139.
18. B. Barboy and B. Gelbart, *Hard-particle fluids. II. General y-Expansion-Like Descriptions*, J. Stat. Phys. **22** (1980), 709.
19. J.-L. Barrat and J.-P. Hansen, *Basic concepts for simple and complex fluids* (CUP, Cambridge, 2003).
20. D. Chandler, *Introduction to modern statistical mechanics* (OUP, Oxford, 1987).

Chapter 5
Maier–Saupe Theory

Abstract In this chapter, we show that the free energy functional introduced in Chap. 4 can be made manageable for thermotropic nematics. We do this by insisting that the molecular field a molecule experiences from the other particles obeys the underlying uniaxial symmetry of the nematic phase and that its magnitude should be proportional to the anisotropy of the van der Waals attraction. By minimising the free energy, we obtain the well-known Maier–Saupe theory for thermotropic nematics. The isotropic–nematic transition temperature we pinpoint by insisting on equal temperatures, number densities of the particles, and equal free energies per unit volume in the co-existing phases, a procedure valid for (nearly) incompressible liquids.

Onsager theory is at the heart of our understanding of lyotropic nematics [1]. In principle, it should be accurate for infinitely rigid, infinitely long rod-like particles interacting via an infinitely harsh repulsive potential. Temperature plays no rôle in Onsager theory. Not surprisingly, this theory has not quite caught on in the field of thermotropic liquid crystals, where the molecules have an aspect ratio of, say, five and where temperature does play a pivotal rôle in directing or suppressing order. Indeed, in thermotropics, the main driving force for the spontaneous alignment of the molecules is not entropy but energy, or, rather, enthalpy, for experiments are usually done at fixed (atmospheric) pressure. For fluids, however, the distinction between energy and enthalpy is of theoretical interest only because these are for all intents and purposes incompressible.

Setting up a theory for thermotropic nematics of comparatively low molecular weight molecules is no trivial matter because these are dense fluids. This implies that a virial expansion is of not much use for this type of liquid crystal, and we need to resort to physical insight to make headway. Taking as starting point the familiar expression that we also used to estimate the excess free energy of a lyotropic system, Eq. (3.5), we have for the potential energy,

$$U_{\text{pot}} = \frac{1}{2} N \int d\vec{u}\, P(\vec{u}) U_{\text{mol}}(\vec{u}),$$

in terms of the molecular field, U_{mol}, where the factor of one-half again corrects for double counting, and where, according to Eq. (3.8), the molecular field

$$U_{\text{mol}}(\vec{u}) = \rho \int d\vec{u}'\, P(\vec{u}') \int d\vec{r}\, g(\vec{r}, \vec{u}, \vec{u}') U(\vec{r}, \vec{u}, \vec{u}')$$

is expressed in terms of the pair correlation function $g(\vec{r}, \vec{u}, \vec{u}')$. As before, $U(\vec{r}, \vec{u}, \vec{u}')$ denotes the pair potential between two molecules of orientations $\vec{u}$ and $\vec{u}'$. Of these two particles, one sits at the origin of the co-ordinate system and the other at position $\vec{r}$. See Fig. 3.2.

The problem that we face now is that the pair correlation function $g(\vec{r}, \vec{u}, \vec{u}')$ is impossible to guess or deduce with any degree of accuracy for dense liquids and that our earlier estimate in terms of a Boltzmann weight of the interaction potential U fails miserably. Hence, we shall not attempt to estimate the pair correlation function itself but focus our attention instead on the molecular field $U_{\text{mol}}(\vec{u})$. Within mean-field theory for classical particles, we expect the orientational distribution function $P(\vec{u})$ to be Boltzmann-weighted, in other words, to be proportional to $\exp(-\beta U_{\text{mol}}(\vec{u}))$. This means that $U_{\text{mol}}(\vec{u})$ must obey the same cylindrical and inversion symmetry properties that the probability P obeys in the nematic phase.

In the isotropic phase, we expect $U_{\text{mol}}(\vec{u}) = U_{\text{iso}}$ to be independent of the orientation $\vec{u}$ of a particle. A *plausible* form of the molecular field in the isotropic and nematic phases must therefore be [2, 3]

$$U_{\text{mol}}(\vec{u}) = U_{\text{iso}} - \Delta\epsilon S P_2(\vec{n} \cdot \vec{u}), \tag{5.1}$$

where $\vec{n}$ denotes as before the director, the second Legendre polynomial $P_2(x) \equiv 3x^2/2 - 1/2$ imposes the correct symmetry, and $S = \langle P_2(\vec{n}\cdot\vec{u}) \rangle = \int d\vec{u}\, P(\vec{u}) P_2(\vec{n}\cdot\vec{u})$ is as before the (scalar) nematic order parameter. See also Fig. 5.1. Recall from Chap. 3 that the nematic order parameter S acquires the value of zero in the isotropic phase, in which case the anisotropic part of the molecular field vanishes. This, of course, is what is required. Finally, $\Delta\epsilon \geq 0$ is a measure of the anisotropy of the attractive van der Waals interaction between the nematogens, favouring parallel orientations. See Fig. 5.2.

Plausibly, U_{iso} and $\Delta\epsilon$ must be some function of the density ρ of the material, as it sets the average distance between the molecules, and through that also on the temperature on account of the thermal expansion of the fluid with increasing temperature. What the precise functional dependence of these quantities on ρ and T is, we can only guess. Within a van der Waals picture, we would expect that $\Delta\epsilon \propto \rho$ [4]. On the other hand, it has been argued that since the van der Waals

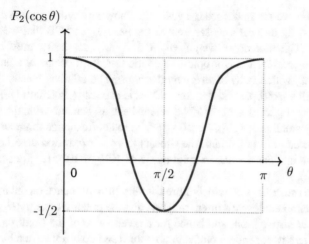

Fig. 5.1 Schematic of the second Legendre polynomial $P_2(\cos\theta)$ as a function of the polar angle θ. Vertical dotted lines indicate angles of symmetry.

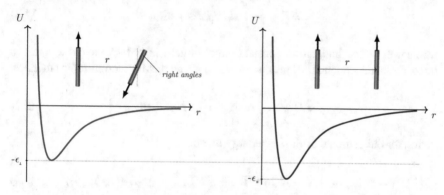

Fig. 5.2 Schematic of the interaction potential U between two nematogens as a function of their centre-to-centre distance r. Left: for two particles at right angles, the potential well is $-\epsilon_\perp$. Right: for parallel particles, the well is deeper, $-\epsilon_\parallel < -\epsilon_\perp$, favouring parallel alignment [6]. The quantity $\Delta\epsilon$ in the molecular field, Eq. (5.1), measures the anisotropy of the van der Waals interaction and is proportional to $\epsilon_\parallel - \epsilon_\perp$.

attraction between molecules must be proportional to r^{-6}, at least for distances large compared to their size, on dimensional grounds, one can also justify the proportionality $\Delta\epsilon \propto \rho^2$.[1] Fortunately, we need not know how U_{iso} and $\Delta\epsilon$ depend on ρ and T to make headway.

---

[1] See the brief overview paper of Stephen Picken and references cited therein [5].

To see how we can make headway without knowing anything much *a priori* about U_{iso} and $\Delta\epsilon$, let us take a closer look at the nematogen 5CB that we encountered in Chap. 3. The molecular weight of 5CB is $M_w \simeq 250\,g\,mol^{-1}$, and under atmospheric pressure the isotropic-to-nematic transition occurs at a temperature of $T_{IN} \simeq 35\,°C$, which is its *clearing temperature*. The relative density difference of the co-existing isotropic and nematic phases is exceedingly small: $(\rho_N - \rho_I)/\rho_I \simeq 0.002$, where $\rho_I \simeq 1\,g\,cm^{-3}$. The latent heat associated with the transition is surprisingly small too: $\Delta H_{IN} \simeq 1.6\,J\,g^{-1}$, which corresponds to about $0.15\,k_B T_{IN}$. Finally, like any kind of liquid, the (linear) thermal expansion co-efficient of 5CB is very small and has a value in the range $\alpha \simeq 10^{-3} - 10^{-4}\,K^{-1}$ depending on the temperature.

All of this suggests that temperature has very little influence on the mean distance between the molecules. We infer from this that the potential energy U_{pot} for a *given orientational distribution*, and hence for a given value of the order parameter S, is only very weakly dependent on the temperature and does not have a hidden entropy component such as is the case for lyotropic systems. Consequently,

$$\beta F_{exc} = \int d\beta U_{pot} \approx \beta U_{pot}, \tag{5.2}$$

where we set the integration constant equal to zero in order to avoid a spurious temperature dependence.[2] Hence, we must conclude that for thermotropic nematics,

$$F_{exc} = U_{pot} = \frac{1}{2}N U_{iso} - \frac{1}{2}N\Delta\epsilon \langle P_2(\vec{n} \cdot \vec{u})\rangle^2, \tag{5.3}$$

giving for our estimate of the free energy functional

$$\beta F[P] = N \ln \rho\omega - N + N\langle \ln P(\vec{u})\rangle + \frac{1}{2}N U_{iso} - \frac{1}{2}N\Delta\epsilon \langle P_2(\vec{n} \cdot \vec{u})\rangle^2. \tag{5.4}$$

It proves convenient to subtract from this free energy the free energy of the corresponding isotropic phase, for which $P = 1/4\pi$. This leaves us with an excess free energy

$$\beta\Delta F[P] \equiv \beta F[P] - \beta F[\frac{1}{4\pi}] = N\langle \ln 4\pi P(\vec{u})\rangle - \frac{1}{2}N\beta\Delta\epsilon \langle P_2(\vec{n} \cdot \vec{u})\rangle^2. \tag{5.5}$$

Notice that we have constructed this free energy in such a way that in the isotropic phase we have $\Delta F \equiv 0$.

The unknown quantity in this expression is the orientational distribution function $P(\vec{u})$. We follow that same recipe as that in the preceding Chap. 5 and functionally

---

[2] This is allowed because the free energy is a *constrained* free energy, that is, defined for a given $P(\vec{u})$. Only after minimisation, $P(\vec{u})$ becomes a function of temperature, not before.

minimise ΔF with respect to $P(\vec{u})$ while insisting that $P(\vec{u})$ remains normalised. Hence, we set

$$\frac{\delta}{\delta P(\vec{u})} \left[\frac{\beta \Delta F}{N} - \lambda \left(\int d\vec{u} \, P(\vec{u}) - 1 \right) \right] = 0, \tag{5.6}$$

where λ is again a Lagrange multiplier that we use to normalise the distribution function. The number of particles N appears for convenience—it only renormalises the Lagrange multiplier. To perform the functional minimisation, we make use of Eq. (4.10) for the first term in the free energy functional, Eq. (5.5). For the second term, we need to realise that

$$\langle P_2(\vec{n} \cdot \vec{u}) \rangle^2 = \int d\vec{u} \int d\vec{u}' \, P(\vec{u}) P(\vec{u}') P_2(\vec{n} \cdot \vec{u}) P_2(\vec{n} \cdot \vec{u}'). \tag{5.7}$$

Inserting $P(\vec{u}) + \delta P(\vec{u})$ and $P(\vec{u}') + \delta P(\vec{u}')$, keeping terms to linear order in $\delta P(\vec{u})$ and a change of variables $\vec{u}' \to \vec{u}$ and $\vec{u} \to \vec{u}'$ finally gives

$$\ln 4\pi P(\vec{u}) + 1 - \beta \Delta \epsilon S P_2(\vec{n} \cdot \vec{u}) - \lambda = 0, \tag{5.8}$$

which again is a self-consistent field equation because $S = \langle P_2(\vec{n} \cdot \vec{u}) \rangle = \int d\vec{u} \, P(\vec{u}) P_2(\vec{n} \cdot \vec{u})$ is a function of the orientational distribution function P that we set out to calculate. Not surprisingly, the structure of the self-consistent field equation obtained within the Maier–Saupe approach closely resembles that of the Onsager equation, Eq. (4.12).

We need to solve this equation to find the orientational distribution function P as a function of the control parameter $\beta \Delta \epsilon$. Noting that Eq. (5.8) can be reformulated in the more familiar form of a Boltzmann distribution, we write

$$P(\vec{u}) = \frac{1}{4\pi} \exp\left(\lambda - 1 + \beta \Delta \epsilon S P_2(\vec{n} \cdot \vec{u}) \right). \tag{5.9}$$

Here, λ can be fixed by demanding that $\int d\vec{u} \, P(\vec{u}) \equiv 1$. We immediately realise that $-\Delta \epsilon S P_2(\vec{n} \cdot \vec{u}) \equiv U_{\text{mol}}(\vec{u})$ now acts as the molecular field that in the isotropic phase, by definition, is zero because there the order parameter is equal to zero, $S = 0$. Hence,

$$P(\vec{u}) = \frac{\exp(\beta \Delta \epsilon S P_2(\vec{n} \cdot \vec{u}))}{\int d\vec{u} \, \exp(\beta \Delta \epsilon S P_2(\vec{n} \cdot \vec{u}))}. \tag{5.10}$$

The order parameter $S = \langle P_2(\vec{n} \cdot \vec{u}) \rangle$ can now be determined self-consistently because using Eq. (5.10) we deduce that

$$S = \frac{\int d\vec{u} \, P_2(\vec{n} \cdot \vec{u}) \exp\left(\beta \Delta \epsilon S P_2(\vec{n} \cdot \vec{u}) \right)}{\int d\vec{u} \, \exp\left(\beta \Delta \epsilon S P_2(\vec{n} \cdot \vec{u}) \right)} \tag{5.11}$$

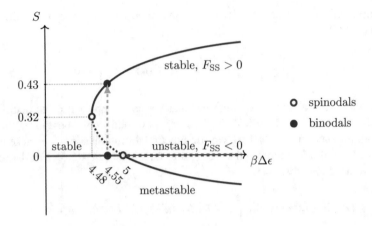

Fig. 5.3 Schematic of the nematic order parameter S as a function of the reciprocal dimensionless temperature $\beta\Delta\epsilon$ for a thermotropic nematic, according to Maier–Saupe theory. Dashed lines indicate thermodynamically unstable branches. The spinodal points are also indicated, as are the binodals. The sign of the second derivative of the free energy F_{SS} with respect to the order parameters is also indicated.

must hold, which has as trivial solution $S = 0$ for all values of $\beta\Delta\epsilon$ as may be verified straightforwardly.

Analytical integration of the self-consistent field equation in closed form has eluded us to date albeit that a complicated expression involving a so-called *special function* can be obtained for $\beta\Delta\epsilon$ in terms of S. (See exercise 2 at the end of this chapter.) Hence, we again take recourse to a numerical evaluation of it. Figure 5.3 shows a schematic of the order parameter S as a function of the dimensionless reciprocal temperature $\beta\Delta\epsilon = \Delta\epsilon/k_BT$. For $\beta\Delta\epsilon < 4.48$, only the isotropic solution with $S = 0$ is stable. For $\beta\Delta\epsilon \geq 4.48$, there are two stable branches, with $\partial^2\Delta F/\partial S^2 > 0$, and one unstable branch, for which $\partial^2\Delta F/\partial S^2 < 0$.

Associated with the solutions are two spinodal points: one at $\beta\Delta\epsilon = 4.48$ with order parameter value $S = 0.32$, which we identify as the nematic spinodal, and one at $\beta\Delta\epsilon = 5$ with order parameter $S = 0$, being the isotropic spinodal. Note that we are again dealing with a universal curve and hence a law of corresponding states. This is because the order parameter S is expressed in terms of a dimensionless temperature scale, $T/k_B\Delta\epsilon$. The shape of the bifurcation curve near the nematic spinodal has the same structure as the binodal of the gas–liquid transition near the critical point and is characterised by a mean-field critical exponent of one-half. Recall that we drew the same conclusion from the Onsager theory, except that in that case we expect the critical exponent to be exact.

To calculate the binodal, and locate the isotropic–nematic transition temperature, we again invoke the second law of thermodynamics, telling us that we must have equal chemical potentials in the co-existing isotropic (I) and nematic (N) phases, $\mu_I = \mu_N$, equal temperatures, $T_I = T_N$, and equal pressures $p_I = p_N$. Unfortunately, we cannot calculate those from our theory, as we do not know how

$\Delta\epsilon$ depends on the density of the fluid. Fortunately, we can make use of the (near) incompressibility property of fluids that we used to formulate the theory in order to work around this problem.

From thermodynamics, we know that the Helmholtz free energy, F, describes a closed system with a fixed number of particles N, a fixed volume V, and a fixed temperature T. The Gibbs free energy, on the other hand, describes a closed system in mechanical equilibrium with a reservoir. This is known as an (N, p, T) ensemble. These two thermodynamic potentials are connected via the Legendre transform $G = F + pV$. In equilibrium, thermodynamics stipulates that $G = N\mu$, with μ the chemical potential of the particles. Hence, $\mu\rho = P + F/V$ with $\rho = N/V$ their number density. From this, we conclude that equal densities, chemical potentials, and pressures translate to equal free energy densities, that is, free energies per unit volume! So, for incompressible systems, the conditions $\mu_I = \mu_N$, $T_I = T_N$, and $p_I = p_N$ may be replaced by $\rho_I = \rho_N$, $T_I = T_N$, and $F_I/V = F_N/V$. The latter equality translates to the condition $\Delta F/V = (F_N - F_I)/V = 0$.

So, it suffices to find the dimensionless reciprocal temperature $\beta_{IN}\Delta\epsilon$ for which $\beta\Delta F/N = 0$. We numerically find that this happens if $\beta_{IN}\Delta\epsilon = 4.55$ and obtain for the order parameter in the nematic phase a value of $S_{IN} = 0.43$. The isotropic and nematic spinodals we locate by evaluating $\partial^2 F/\partial S^2 = F_{SS} = 0$, to give $\beta_I^{\text{spinodal}}\Delta\epsilon = 5$ with $S_I = 0$, and $\beta_I^{\text{spinodal}}\Delta\epsilon = 4.48$ and $S_N = 0.32$. These values are indicated in Fig. 5.3, as are the stable, metastable, and unstable solutions. See also exercises 2 and 3 at the end of this chapter. The branch with negative order parameter is metastable, so has a free energy larger than that of positive order parameter at equal temperature.

Extracting energy from the isotropic phase reduces its temperature until we reach the clearing temperature for which $\beta_{IN}\Delta\epsilon = 4.55$. Extracting more energy does not lead to a further decrease in temperature. What happens instead is that a part of the isotropic phase is converted into the nematic phase, in proportion to how much of the latent heat has been extracted from the fluid. See Fig. 5.4. Once all the latent

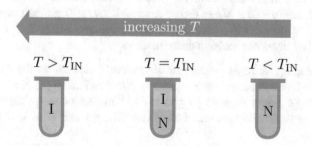

Fig. 5.4 Schematic representation of how isotropic and nematic phases present themselves in a thermotropic system. Left: for temperatures above where the nematic phase emerges, we observe only the isotropic phase. Right: for temperatures below the clearing temperature again only one phase is seen, which is the nematic phase. Middle: if not all of the latent heat is removed from the isotropic fluid, we have co-existence of an isotropic phase and a nematic phase.

heat has been extracted, the temperature decreases again and the order parameter increases. Notice that a nematic with negative order parameter is metastable: its free energy is larger than that with positive order parameter at equal temperature. Even if we could somehow prepare a nematic fluid with negative order parameter, it would still not be long-lived.

The latent heat per particle that we obtain from Maier–Saupe theory equals $\Delta U_{\mathrm{IN}}/N = \partial \beta \Delta F / \partial \beta N|_{T=T_{\mathrm{IN}}} = 0.42\ k_{\mathrm{B}} T_{\mathrm{IN}}$. It tells us that, according to Maier–Saupe theory, each particle gains an amount just shy of half a thermal energy through collective alignment at the I-N transition, which exactly compensates for the loss of orientational entropy. Notice that this value is larger than the measured value of $0.15\ k_{\mathrm{B}} T_{\mathrm{IN}}$ for 5CB. The temperature dependence of measured order parameters reasonably accurately follows the prediction of Maier–Saupe theory, presuming that $\Delta \epsilon$ does not depend on temperature itself, which was one of the assumptions of our derivation. In general, Maier–Saupe theory is considered as successful, despite its simplicity and obvious drawbacks [7].

One drawback, the lack of a (tiny) density jump at the transition, seems not so serious and can be fixed relatively straightforwardly [8]. Another drawback, the mean-field approximation, is more severe. According to the theory, the isotropic spinodal temperature is only 1.5% lower than the clearing temperature, which corresponds to approximately 5 K for 5CB. In reality, the difference is much smaller, about 1 K. This means that near the I-N transition, critical fluctuations due to the nearby spinodal cannot be ignored. Approaching the spinodal, the free energy cost of excitations becomes negligible. This causes the amplitude of spontaneous fluctuations to become very large. (See exercise 1 of this chapter.) These fluctuations drive the transition to become weakly first order, as may in fact also be deduced from the small latent heat [9]. This has important consequences, e.g., for the heat capacity near the isotropic–nematic phase transition, which due to the impact of the fluctuations deviates strongly from that predicted from mean-field theories such as the Maier–Saupe theory.

This concludes our analysis of the Maier–Saupe theory. Next we will dwell on extensions of the Maier–Saupe and the Onsager theory and see what it takes to turn the Onsager theory into a Maier theory (and *vice versa*). This then suggests ways of merging both theories in a phenomenological Onsager–Maier–Saupe model that deals with the effects of repulsion and attraction.

Further Reading A lucid discussion of modern variations on the Maier–Saupe theme can be found in a short review paper by Stephen Picken [5]. A clear overview of thermotropic liquid crystals in general and molecular statistical theories, in particular, can be found in the textbook of Ger Vertogen and Wim de Jeu [10].

Exercises

1. **Spinodal instability of the isotropic phase.** Even without functionally minimising the free energy of the Maier–Saupe theory, Eq. (5.5), provides information about the statistics of the isotropic phase. To probe the stability of the isotropic phase, we presume a weak nematic perturbation of the orientation distribution function. We write $P(\vec{u}) = P(x)/2\pi$ with $\int_{-1}^{+1} dx\, P(x) = 1$ and use as *ansatz* or *trial function*:

$$P(x) = \frac{1}{2} + \varepsilon P_2(x), \tag{5.12}$$

where $x = \vec{n} \cdot \vec{u} = \cos\theta$ in terms of the polar angle $\cos\theta$ between the director $\vec{n}$ and the main body axis vector $\vec{u}$ of the nematogens. Here, ε is a small parameter, and $P_2(x) = (3x^2 - 1)/2$ the second Legendre polynomial. Note that the distribution function is normalised because $\int d\vec{u}\, P_2(x) = 0$.

(a) Show, by making use of a Taylor expansion, Eq. (1.10), that for $|\varepsilon| \ll 1$ the free energy obeys

$$\beta\Delta F = \beta\Delta F(S) = \frac{5}{2}N\left(1 - \frac{1}{5}\beta\Delta\epsilon\right)S^2 + \cdots \tag{5.13}$$

at least to lowest order in the nematic order parameter, S. Recall from exercise 3 of Chap. 5 that for the given ansatz of the distribution function, Eq. (5.12), we have $S = \langle P_2(x)\rangle = 2\varepsilon/5$.

(b) Explain why this shows that $\beta\Delta\epsilon = 5$ at the isotropic spinodal and that the isotropic phase is only stable if $\beta\Delta\epsilon < 5$.

(c) Show that for $\beta\Delta\epsilon < 5$, a small spontaneous fluctuation to a non-zero value of the order parameter decreases the entropy more than can be compensated for by a lowering of the energy. Hint: deduce which part of the free energy Eq. (5.13) represents energy and which entropy, and presume $\Delta\epsilon$ does not depend on the temperature.

According to statistical mechanical theory, [11], the probability distribution function $P(S)$ of the order nematic parameter S is given by the Boltzmann distribution

$$P(S) = \frac{\exp\left(-\beta\Delta F(S)\right)}{\int_{-1/2}^{+1} dS \exp\left(-\beta\Delta F(S)\right)}. \tag{5.14}$$

This we can now use to calculate the expectation value and variance of the order parameter.

(d) To be able to do this, we can make use of the correspondence of Eq. (5.14) to the Gaussian distribution for some stochastic variable $x \in (-\infty, +\infty)$,

$$P(x) = \frac{1}{\sqrt{4\pi \sigma_x^2}} \exp\left(-\frac{(x - \langle x \rangle)^2}{2\sigma_x^2}\right), \tag{5.15}$$

with $\langle x \rangle = \int_{-\infty}^{+\infty} dx\, x\, P(x)$ the expectation value of x, and σ_x^2 its variance. Verify that this distribution function is normalised and that $\sigma_x^2 = \langle (x - \langle x \rangle)^2 \rangle = \langle x^2 \rangle - \langle x \rangle^2$ as it should. Note that $\int_{-\infty}^{+\infty} dt \exp -t^2 = \sqrt{\pi}$.

(e) Show, using the correspondence to the Gaussian distribution function in the thermodynamic limit $N \gg 1$, that the expectation value of the order parameter is that of the isotropic phase, $\langle S \rangle = 0$, provided $\beta\Delta\epsilon < 5$, and that the variance of the order parameter,

$$\sigma_S^2 = \langle S^2 \rangle - \langle S \rangle^2 = \frac{1}{5N\left(1 - \frac{1}{5}\beta\Delta\epsilon\right)}, \tag{5.16}$$

diverges upon approaching the spinodal. What does this divergence signify?

2. **Solving the Maier–Saupe equation.** The self-consistent field equation, Eq. (5.11), of the Maier–Saupe model produces the expectation value of the order parameter $S \equiv \langle S \rangle$ as a function of the energetic parameter $\beta\Delta\epsilon$, shown in Fig. 5.3.

(a) Verify that this self-consistent field equation is equivalent to the expression $S = S(K) = d\ln Z/dK$ with $Z = Z(K) = \int d\vec{u} \exp\left[K P_2(\vec{n} \cdot \vec{u})\right]$ a single-particle partition function and $K \equiv \beta\Delta\epsilon S$.

(b) Show that the single-particle partition function reads

$$Z(K) = \frac{2\pi^{3/2}}{\sqrt{\frac{3}{2}K}} \exp\left(-\frac{1}{2}K\right) \mathrm{erfi}\left(\sqrt{\frac{3}{2}K}\right) \tag{5.17}$$

in terms of the *imaginary error function* $\mathrm{erfi}(z) = -i\,\mathrm{erf}(iz)$, with $i = \sqrt{-1}$ the imaginary number, z a (complex) variable, and $\mathrm{erf}(z) = 2\pi^{-1/2} \int_0^z dt \exp(-t^2)$ the usual *error function* [12]. This shows that a simple analytical expression for the order parameter as a function of the temperature (or *vice versa!*) is not straightforward to obtain for the Maier–Saupe theory.

(c) We can rewrite the self-consistent field equation of a) as $\beta\Delta\epsilon = K/(d\ln Z/dK)$, while $S = K/\beta\Delta\epsilon$. Consequently, we can take K as a "dummy" variable and calculate $\beta\Delta\epsilon$ as a function of S numerically for a range of values of $K \in [-1.5, 5]$. Do this using your favourite mathematical software package, and check that for $\beta\Delta\epsilon = 5$ non-trivial solutions with positive and negative order parameter branch off from the trivial solution.

(d) Calculate the free energy $\beta \Delta F/N$ as a function of $-1/2 \le S \le 1$ for different values of $\beta \Delta \epsilon$, and verify the locations of the spinodal and binodal points mentioned in the main text. Hint: verify first that $\beta \Delta F/N = -\ln(Z/2) + \beta \Delta \epsilon S^2/2$ if we take out the integration over the azimuthal angle and redefine $Z \equiv Z/2\pi$.

3. **The nematic–smectic A transition.** Some compounds that exhibit a nematic phase also exhibit a smectic A phase under conditions in between those where the nematic and the crystal phases are stable. The smectic A phase resembles the nematic in that the particles that make up the fluid are aligned along a director but differ in the sense that the particle density is not uniform but modulated along the director. In a way, particles attain crystalline order along the director but remain fluid-like perpendicular to that. See Fig. 2.1. The typical wavelength of the density modulation is the size of the molecules, d, and the density variation is proportional to $\cos(2\pi z/d)$ with z the spatial co-ordinate along the director. A simplified version of the theory proposed William McMillan in the 1970s, [13], in which we ignore the influence of differences in degree of orientational order and that has the same ingredients as that of Wilhelm Maier and Alfred Saupe describing the nematic transition in thermotropics, is based on the following Helmholtz free energy functional

$$\frac{\beta \Delta F[\psi]}{N} = \int_{-d/2}^{+d/2} dz \left[\psi(z) \ln(\psi(z)d) + \frac{1}{2} \psi(z) \beta \Delta U_{\text{mol}}(z) \right]. \quad (5.18)$$

Here, N is the number of particles in a slab of thickness d, reflecting the periodicity of the density modulation, $\psi(z)$ is the positional distribution function of the centres of mass of the particles in the slab, and ΔU_{mol} is the excess molecular field driving the nematic transition. The distribution function is normalised, $\int_{-d/2}^{+d/2} dz \, \psi(z) = 1$. Further, $\beta = 1/k_B T$ is as usual the reciprocal thermal energy, with k_B Boltzmann's constant and T the absolute temperature.

(a) Describe the physical background of the two ingredients in this free energy functional, and explain what simplifying assumption has been made.

A sensible trial function for the molecular field is

$$\Delta U_{\text{mol}}(z) = \Delta \epsilon \sigma \cos \left(\frac{2\pi z}{d} \right), \quad (5.19)$$

where $\Delta \epsilon$ is the strength of the van der Waals interaction between the particles driving the transition to the smectic A phase. Arguably, this is strongest for particles in register, that is, for particles that line up perfectly in a layer. The degree of positional ordering is described by the smectic order parameter σ, defined as

$$\sigma = \int_{-d/2}^{+d/2} dz \, \psi(z) \cos \left(\frac{2\pi z}{d} \right). \quad (5.20)$$

(b) Show that the most probable distribution of material in the smectic A phase obeys the following self-consistent field equation,

$$\psi(z) = \frac{1}{Z} \exp\left[+\beta\Delta\epsilon\sigma \cos\left(\frac{2\pi z}{d}\right)\right], \tag{5.21}$$

where Z acts as a normalisation constant.

(c) Show that the uniform distribution describing the nematic phase, with smectic order parameter $\sigma = 0$, is a solution to Eqs. (5.20) and (5.21) for all temperatures T, and that the associated free energy obeys $\Delta F = 0$, implying that the free energy functional describes the difference in free energies of the smectic phase and the reference nematic phase.

The self-consistent field equation can be solved by presuming the ordering is weak and $\sigma \ll 1$. In that case, we find for the order parameter a second solution

$$\sigma = \frac{1}{2}\sqrt{\frac{\beta\Delta\epsilon}{4} - 1}, \tag{5.22}$$

which becomes real and hence relevant if $\beta\Delta\epsilon \geq 4$, i.e., for $T \leq \epsilon/4k_B$. This fixes the nematic–smectic A transition temperature at $T_{N\,SmA} = \epsilon/4k_B$. According to the simple model, the transition is continuous, that is, of second order, with a mean-field critical exponent of one-half.

In the original theory of McMillan, which includes a description of the degree of orientational order, the nematic–smectic transition can be first or second order, depending on the system parameters [13]. Experimentally, and theoretically, the order of the nematic–smectic A transition remains a matter of contention and may depend on the material [14].

(d) Describe how we can verify that for $T < T_{N\,SmA}$ the smectic A phase is indeed the thermodynamically stable phase and that the nematic phase must be thermodynamically unstable. Show for this purpose that if $\sigma \ll 1$, we must have

$$\frac{\beta\Delta F}{N} = 2\sigma^2\left(1 - \frac{1}{4}\beta\Delta\epsilon\right) + \cdots, \tag{5.23}$$

if we use as ansatz (trial function) for the distribution $\psi(z) = d^{-1}\left[1 + \varepsilon\cos\frac{2\pi z}{d}\right]$ with ε a small parameter. Make use of the identity $\int_{-d/2}^{+d/2} dz\cos^2(2\pi z/d) = d/2$, to verify Eq. (5.23).

References

1. G. J. Vroege and H. N. W. Lekkerkerker, *Phase transitions in lyotropic colloidal and polymer liquid crystals*, Rep. Prog. Phys. **55** (1992), 1241.
2. W. Maier and A. Saupe, *Eine einfache molekular-statistische Theorie der nematischen kristallinflüssigen Phase. Teil I*, Zeitschrift für Naturforschung A, **14** (1960), 882.
3. W. Maier and A. Saupe, *Eine einfache molekular-statistische Theorie der nematischen kristallinflüssigen Phase. Teil II*, Zeitschrift für Naturforschung A, **15** (1960), 287.
4. M. A. Cotter, *Generalized van der Waals theory of nematic liquid crystals: an alternative formulation*, J. Chem. Phys. **66** (1977), 4710.
5. S. J. Picken, *Orientational order in nematic polymers - some variations on the Maier-Saupe theme*, Liquid Crystals, 37 (2010), 977.
6. V. A. Parsegian, *Van der Waals forces: a handbook for biologists, chemists, engineers, and physicists* (CUP, Cambridge, 2006).
7. S. Chandrasekhar, *Liquid crystals*, 2nd edition (CUP, Cambridge, 1992).
8. Y. Jiang, C. Greco, K. Ch. Daoulas and J. Z. Y. Chen, *Thermodynamics of a compressible Maier-Saupe model based on the self-consistent field theory of wormlike polymer* , Polymers **9** (2017), 48.
9. P. K. Mukherjee, *The $T_{NI} - T^*$ puzzle of the nematic-isotropic phase transition*, J. Phys. Condens. Matter **10** (1998), 9191.
10. G. Vertogen and W. H. de Jeu, *Thermotropic liquid crystals, fundamentals* (Springer, Berlin, 1988).
11. D. Chandler, *Introduction to modern statistical mechanics* (OUP, Oxford, 1987).
12. D. A. McQuarrie, *Mathematical methods for scientists and engineers* (USB, Sausalito, 2003).
13. W. L. McMillan, *Simple molecular model for the smectic A phase of liquid crystals*, Phys. Rev. A **4** (1971), 1238.
14. S. Singh, *Phase transitions in liquid crystals*, Phys. Rep. **324** (2000), 107.

Chapter 6
Beyond Maier–Saupe and Onsager

Abstract In this final chapter, we touch upon extensions of Onsager and Maier–Saupe theory, a Maier–Saupe-type version of Onsager theory, and discuss potential interpolations between Onsager and Maier–Saupe theory.

We have obtained Onsager theory and Maier–Saupe theory using the same density functional formalism by making appropriate approximations that suit the differences in the physics underlying the spontaneous ordering that takes place in lyotropic and thermotropic nematic phases. We recall that lyotropics are dominated by competing forms of entropy, and thermotropics by a competition between enthalpy (or energy) and entropy. The fact that we can obtain both theories from the same root theory shows that the theories are, in a way, two sides of the same coin. Although perhaps with hindsight not terribly surprising, this also explains why the governing integral equations for the orientational distribution functions are so similar.

Both theories are, in spite of their drawbacks, considered central to our understanding of nematic liquid crystals, even though much more sophisticated theories have since been put forward that attempt to deal more accurately with (aspects of) experimental reality. This includes accounting for a finite bending flexibility, a finite aspect ratio, an underlying helical shape or chiral interaction, polydispersity in length and/or width, the presence of other types of colloid or solvent, external fields, the self-assembly of the particles from smaller molecular building blocks, and other types of interactions between them, such as screened electrostatic interactions. The literature is literally awash with theoretical and simulation studies that attempt to deal with these complications.

Let us, by way of example, single out one experimentally relevant aspect, namely the impact of a finite bending flexibility. If cylindrical particles are not infinitely stiff, then thermal agitation will generate configurations of such particles that are not perfectly straight. Indeed, if in some sense sufficiently flexible, the average shape of an elongated particle resembles a coil rather than a rod. This does not preclude the particles from supporting a nematic liquid crystalline phase, as long as local interaction between portions of the coils remains sufficiently directional, i.e., anisotropic. This implies that the particles should not be too flexible either,

© The Author(s), under exclusive license to Springer Nature Switzerland AG 2022
P. van der Schoot, *Molecular Theory of Nematic (and Other) Liquid Crystals*,
SpringerBriefs in Physics, https://doi.org/10.1007/978-3-030-99862-2_6

and hence, we refer to such particles as being *semi-flexible*. The loss of entropy is now associated with the suppression of bending fluctuations in a phase where the particles are aligned. In the nematic phase, the alignment causes the coils to become stretched and lose their spheroidal average shape.

The isotropic-to-nematic phase transition in *dispersions* or *melts* of very long semi-flexible polymers can be described by the same Onsager and Maier–Saupe theories for rigid particles we discussed in the preceding two chapters, simply by replacing the expression for the orientational free energy F_{or}, Eq. (3.4), by Khokhlov and Semenov [1–4]

$$F_{or} = -k_B T N \frac{L}{2L_P} \int d\vec{u} \sqrt{P(\vec{u})} \nabla_{\vec{u}}^2 \sqrt{P(\vec{u})}. \tag{6.1}$$

Here, L again denotes the length of the cylinder, L_P is the so-called *persistence length*, $P(\vec{u})$ as before the orientational distribution function, and $\nabla_{\vec{u}}^2$ the Laplace operator on the unit sphere. In the usual polar co-ordinates θ and φ, and $x \equiv \cos\theta$, this operator attains the form $\nabla_{\vec{u}}^2(\cdots) = \frac{\partial}{\partial x}\left((1-x^2)\frac{\partial}{\partial x}(\cdots)\right) + \frac{1}{1-x^2}\frac{\partial^2}{\partial\varphi^2}(\cdots)$. For liquid crystals with uniaxial symmetry, this simplifies to $\nabla_{\vec{u}}^2(\cdots) \rightarrow \nabla_x^2(\cdots) = \frac{\partial}{\partial x}\left((1-x^2)\frac{\partial}{\partial x}(\cdots)\right)$.

Implied in Eq. (6.1) is the presumption that length L of the cylindrical particle is (very) much larger than its persistence length L_P. The persistence length is a material constant and a measure for the bending stiffness of the particle. It is the mean distance along the curved centre line of the cylinder over which its tangent vector decorrelates, so loses its memory where it came from, due to thermal motion in an isotropic fluid. This molecular length scale depends on the chemistry of the material, the kind of solvent, if there is one, and the temperature. If the length L of the filamentous particle is much smaller than persistence length L_P, its shape is that of a straight rod. On the other hand, if $L \gg L_P$, then, in isotropic solution, its shape resembles that of a coil, with a mean *end-to-end distance* R that obeys $R^2 = 2LL_P$.[1] See Fig. 6.1.

Alexey Khokhlov and Alexander Semenov derived Eq. (6.1) in a *tour de force* from the statistical physics of fluctuating elastic fibres in an external orienting field and argued that the excess free energy for hard, semi-flexible cylinders must, to a good approximation, be the same as that for hard rigid cylinders. This is accurate provided the diameter D of the cylinders is very much smaller than the persistence length L_P, and the interactions are dominated by repulsive interactions. The reason is that the excess free energy density for hard rods scales as

$$\beta F_{exc}/V \propto \rho^2 L^2 D \propto \phi^2 D^{-3} \tag{6.2}$$

to leading order in the aspect ratio $L/D \gg 1$, with ϕ again the volume fraction of the particles, as we have seen in Chap. 5. (See Eqs. (4.3) to (4.6) of Chap. 5.)

---

[1] Strictly speaking, this expression holds if excluded volume interactions within the chain are weak.

Fig. 6.1 Schematic representation of a model configuration for a semi-flexible worm-like particle. L denotes its contour (or extended) length, D its width, and L_P the so-called persistence length, which is a measure for the bending stiffness of the particle.

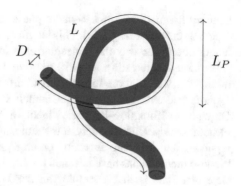

This excess free energy is independent of the aspect ratio of the rods! Consequently, breaking up rods into smaller portions does not change the excess free energy, at least if the shorter rods remain slender and the Onsager limit $L/D \gg 1$ remains to hold. In that case, the second virial approximation remains valid.

A corollary of this observation is that chain connectivity must be irrelevant and that a long semi-flexible chain can be seen as a collection of rod-like segments measuring about a persistence length, at least as far as the excess free energy of the dispersion is concerned. This may seem somewhat counter-intuitive or even highly dubious and in fact absurd. However, from the statistical theory of polymer molecules, we know that once the mean mesh size of a dispersion of entangled polymers drops below about a persistence length, chains cannot distinguish between their own persistence segments and those of others [5]. Good agreement between the predictions from Khokhlov–Semenov theory and experiments lends support the validity of this view, at least in the appropriate limit of $L_P \gg D$ [6].

If we accept this argument in the limit $L_P \gg D$ where it is supposed to hold, the theory shows that for semi-flexible hard chains the relevant concentration scale is not $\phi L/D$ but rather $\phi L_P/D$. Hence, the isotropic-to-nematic transition shifts to higher concentrations, with $\phi_I L_P/D = 5.1$ and $\phi_N L_P/D = 5.5$, *independently* of the chain length L [6]. (See also the exercise 1 below.) Recall that for straight hard rods, $\phi_I L/D = 3.3$ and $\phi_N L/D = 4.2$. That for sufficiently long semi-flexible chains the length no longer plays a significant role is not all that surprising, given that for both in the orientational free energy, Eq. (6.1), and the excess free energy, Eq. (6.2), the chain length can be absorbed in the volume fraction ϕ.

A numerical evaluation of the complete theory valid for arbitrary values of L/L_P shows that the transition crosses over smoothly from that for rigid rods $L/L_P \ll 1$ to that of semi-flexible chains $L/L_P \gg 1$ [7]. Note that strictly speaking, the particles are in the slender-rod limit if $D \ll L \ll L_P$ and in the semi-flexible chain limit if $L \gg L_P \gg D$. Only then the second virial approximation should be expected to hold [7].

In practice, the rigid-rod limit is approached only for very small values of $L_P/L \lesssim 0.01$, while the long-chain limit sets in relatively quickly for $L \gtrsim L_P$ [7]. The reason is that the persistence length is not the appropriate length scale for bending fluctuations in the nematic phase, but a quantity known as the *deflection length* or the *Odijk length* [8]. This length scale is much smaller than the persistence length due to the way the molecular ordering field freezes out shape fluctuations. Even though interesting, we will not dwell on this technicality here and refer to intuitive as well as more formal derivations in the review paper by Theo Odijk[2] [8].

A natural question presents its, namely if we can somehow interpolate between Onsager (or Khokhlov–Semenov) theory and Maier–Saupe theory and get the best of both worlds. Unfortunately, a tractable and accurate theory that treats hard-core repulsion and soft-core attraction on an equal footing remains elusive. This is not because theoreticians have not tried [9–13]. An important reason for this lamentable state of affairs is that a simple van der Waals-type of approximation, in which the impact of van der Waals attractions on the thermodynamics of the fluid of non-spherical particles is treated separately from the hard-core repulsion, does not accurately account for what may be referred to as *translation-orientation coupling*. This is particularly important for slender-rod-like particles and expresses itself in how attractive and repulsive interactions affect the second and higher order virial co-efficients [14]. It turns out that van der Waals interactions may be weak enough not to significantly affect the second virial co-efficient, and remain positive and dominated by a hard-core repulsion, yet still give rise to a third virial co-efficient that is strongly negative and dominated by the attractive interactions [14]. As a consequence, the second virial approximation fails for rigid rods interacting via relatively weak attractive forces, even if they are very slender.[3]

If the van der Waals interactions between rod-like colloidal particles are very weak, to the extent of becoming (in a manner of speaking) "homeopathic" in strength, we may choose to ignore this inconvenience and attempt morphing the Onsager theory into a Maier–Saupe theory. This is actually not such a difficult trick to pull off, if we make use of the *addition theorem* of spherical harmonics. In exercise 3 of Chap. 5, we used it to simplify the kernel of the integral operator in the Onsager equation, Eq. (4.12), producing Eq. (4.21). Insisting on the normalisation of the orientational distribution function, $P(x)$, this integral equation can be expressed in the simple Boltzmann form

$$P(x) = \frac{\exp\left[\frac{5}{4}cSP_2(x)\right]}{\int_{-1}^{+1} dx \exp\left[\frac{5}{4}cSP_2(x)\right]},\tag{6.3}$$

where $S = \langle P_2(x) \rangle$ is, as before, the scalar nematic order parameter, and $c = \phi L/D$ the appropriately scaled concentration. Comparing this with the expression for the orientational distribution function with the one from Maier–Saupe theory, Eq. (5.11), shows that we now have in our hands a mapping of the Onsager theory

---

[2] Equation (6.1) suggests a natural length scale proportional to $P / \int d\vec{u} \sqrt{P(\vec{u})} \nabla_u^2 \sqrt{P(\vec{u})}$.

[3] A finite bending flexibility may actually come to the rescue here, as bending fluctuations give rise to a long-range repulsion that arguably weaken the effect of van der Waals interactions [15].

onto the Maier–Saupe theory.[4] The mapping is simple: $\beta\Delta\epsilon \rightarrow 5c/4$. Using this mapping turns the spinodal for the isotropic phase in the Maier–Saupe theory, $\beta\Delta\epsilon = 5$, into the correct one for the Onsager theory, namely, $c = 4$.

According to Maier–Saupe theory, which presumes equal densities of particles in the co-existing phases, the isotropic-to-nematic transition occurs for the thermotropic nematogens under conditions where $\beta_{IN}\Delta\epsilon \simeq 4.6$. If the mapping is exact, this then suggests that the corresponding transition for lyotropic nematogens should occur at $c_{IN} \simeq 3.6$. This, of course, is nonsensical, for we have seen that the concentrations in the co-existing isotropic and nematic phases in lyotropic nematics are different, the nematic phase being the more concentrated one. This is the main difference between the Maier–Saupe theory and the Onsager theory: the former presumes incompressibility of the fluid and the latter does not, so the mapping is incomplete.

What we need to do to accommodate the difference between thermotropic and lyotropic nematics is to explicitly equate the chemical potentials and osmotic pressures in the co-existing isotropic and nematic phases. (See exercise 2 of Chap. 5.) If we (1) numerically solve the self-consistent field equation $S = \langle P_2(x) \rangle$ for the Maier–Saupe version of the hard-rod distribution function, Eq. (6.3), and (2) use that to calculate the chemical potentials and osmotic pressures for the isotropic and nematic phases, we find $c_I \simeq 3.5$, $c_N \simeq 3.9$ and an order parameter of $S \simeq 0.6$ in the co-existing nematic phase [16]. This compares reasonably well with what we found using the full Onsager theory, $c_I \simeq 3.3$, $c_N \simeq 4.2$, and an order parameter of $S \simeq 0.8$.

This now suggests an obvious way of merging Onsager theory and Maier–Saupe theory, at least for elongated rod-like particles that are not only hard but also a little sticky on account of the presence of residual van der Waals interactions acting between them. Indeed, if we replace $5c/4$ by $\beta\Delta\epsilon + 5c/4$ in Eq. (6.3), we should be done. Actually, this is not quite the case because we now need to worry about the concentration dependence of the energetic parameter $\Delta\epsilon$. In the spirit of van der Waals theory, we would expect $\beta\Delta\epsilon \propto c$ because the molecular field every particle experiences must be proportional to the density, as we discussed in Chap. 6[5]. If we accept this *ansatz*, we are in actual fact treating the hard-core repulsion and the van der Waals attraction within a second virial theory, which, as already mentioned, would only be accurate for hard rods that in addition interact via a very weak attraction between them.

Within this prescription, it makes sense to write $\Delta\epsilon \equiv 5\Delta\epsilon'c/4$, with $\Delta\epsilon'$ some pair contact energy that merely renormalises the dimensionless concentration scale. We can now define as the relevant concentration scale $d \equiv c(1 + \beta\Delta\epsilon') = c(1 + \Delta\epsilon'/k_BT)$ and conclude that according to our Onsager–Maier–Saupe theory, we must have $d_I \simeq 3.5$, $d_N \simeq 3.9$ and $S \simeq 0.6$ under conditions of phase co-existence.

---

[4] It has also been dubbed the "P_2 Onsager theory" [16].

[5] Maier and Saupe suggest that $\beta\Delta\epsilon \propto c^2$ on account of the $1/r^6$ dependence of the van der Waals interaction [17, 18].

This implies that the concentrations in the co-existing phases shift to lower values the lower the temperature, because $c_I \simeq 3.5/(1 + \Delta\epsilon'/k_B T)$ and $c_N \simeq 3.9/(1 + \Delta\epsilon'/k_B T)$. This is not surprising, of course, and supported by experimental data [12]. What we do *not* find is the sudden widening of the phase gap below some temperature found experimentally, [19] and theoretically within van der Waals-type of approximations [1, 20], leading to co-existence between a very dilute isotropic phase and a dense nematic with very little solvent. See also Fig. 2.9.

Considering how differently we obtained the Onsager and Maier–Saupe theories from the underlying density functional formalism, we are forced to conclude that a simplistic interpolation, such as the one sketched above, will just not do. This is not to say that creative phenomenological interpolations are not possible that even have some merit [12, 13, 19]. It does mean that thermodynamic consistency is difficult to enforce and arguably that a different route has to be found for this purpose. Of course, one could also be pragmatic and construct a generalised van der Waals theory, the beginnings of which we sketched in exercise 3 of Chap. 3 or rely on lattice-statistics-based theories, such as that of Paul Flory, where the use is made of a fluid lattice model to describe the configurations of the particles [21]. All of this is outside of the scope of this introductory exposition.

Finally, both Onsager and Maier–Saupe theories have been extended in the literature to describe transitions to chiral nematic phases, various smectic phases, columnar phases, and so on. Arguably, the denser the phase, the more contentious the various assumptions and approximations become, and, also, the more difficult computer simulations become, in particular if the particles are much longer than that they are wide, that is, have a large aspect ratio. We encourage the interested readers to study theoretical papers on this topic and see how these connect with the presentation offered in this monograph.

Further Reading Few review papers or even books discuss the full breadth of applications of Onsager-like and Maier–Saupe-like theories of liquid crystals, albeit that the review paper of Shri Singh provides perhaps the most diverse overview [22]. A detailed exposition of Maier–Saupe theories for thermotropic nematics, and smectics, including chiral versions of these phases, can be found in the book of Ger Vertogen and Wim de Jeu [23]. Arguably, the most extensive review on lyotropic liquid crystals remains that of Gert-Jan Vroege and Henk Lekkerkerker [6]. A more recent one with a stronger focus on experiments is that by Zvonimir Dogic and Seth Fraden [24]. A rather more extensive (and technical) review of density functional theory and simulations of hard convex bodies, which includes a discussion of biaxial nematic, smectic, and columnar phases, is that by Mike Allen et al. [25]. The textbook of Remco Tuinier and Henk Lekkerkerker discusses, amongst other things, extensions of Onsager theory for short rods and what happens if rod-like particles are mixed with spherical colloidal particles [26]. An overview of recent work on liquid crystals and the emergence of "hyper complexity" in them is that by Zvonimir Dogic et al. [27].

Exercises

(1) **Lyotropic nematics of semi-flexible chains.** Khokhlov–Semenov theory was originally derived for lyotropic nematics of semi-flexible cylindrical particles dispersed in a solvent. The particles are presumed to interact via a hard-core repulsion.

(a) Make plausible that within Khokhlov–Semenov theory for lyotropic nematics in the limit $L \gg L_P \gg D$, the relevant concentration scale *must* be $\rho L L_P D$, by comparing the most relevant terms in the overall free energy, Eq. (3.1), replacing the orientational free energy Eq. (3.4) by Eq. (6.1).

(b) Because $P(\vec{u}) = P(x)/2\pi$ with $x = \cos\theta$ is a function only of the polar angle θ, the Laplace operator $\nabla_{\vec{u}}^2(\cdots)$ can be simplified to

$$\nabla_x^2(\cdots) \equiv \frac{\partial}{\partial x}\left((1 - x^2)\frac{\partial}{\partial x}(\cdots)\right). \tag{6.4}$$

Verify that for real integrable functions $f(x)$ and $g(x)$ on the domain $x \in [-1, +1]$, the equality $\int_{-1}^{+1} dx f(x)\nabla_x^2 g(x) = \int_{-1}^{+1} dx g(x)\nabla_x^2 g(x)$ holds. This means that the Laplace operator ∇_x^2 is *self-adjoint*, that is, *Hermitian* [28].

(c) Demonstrate by functionally minimising the free energy that the equivalent of the Onsager equation for the Khokhlov–Semenov theory must read

$$\frac{1}{\sqrt{P(x)}}\nabla_x^2\sqrt{P(x)} = -\lambda + \frac{8c_p}{\pi^2}\langle|\sin\gamma|\rangle', \tag{6.5}$$

where $c_p = \phi L_P/D$ is a scaled concentration, and λ a Lagrange multiplier enforcing the normalisation condition of the distribution function $P(x)$. Furthermore, $\phi = \pi\rho L D^2/4$ is the volume fraction of cylindrical particles and $\langle\cdots\rangle \equiv \int d\vec{u} P(x)(\cdots)$ now denotes an orientation average that already accounts for the uniform distribution in the azimuthal angle.

(d) Show that $P(x) = P = \lambda/8c_p = 1/2$ is the trivial solution of the integro-differential equation, Eq. (6.5). Use Eq. (4.20) to evaluate the molecular field contribution.

(e) Show that a nematic solution of Eq. (6.5) branches off from the trivial one for $c_p = 6$. This turns out to be the isotropic spinodal. Hint: Review exercise 3 of Chap. 6. Make use of the fact that the Legendre polynomials are eigen functions of the operator ∇_x^2, with $\nabla_x^2 P_2(x) = -6P_2(x)$, and recall that $\int_{-1}^{+1} dx P_2(x) = 0$, and $\int_{-1}^{+1} dx P_2^2(x) = 2/5$.

(2) **Thermotropic nematics of semi-flexible chains.** Khokhlov–Semenov theory can straightforwardly be applied to thermotropic nematics. We focus on the limit $L \gg L_P$.

(a) Demonstrate that a Maier–Saupe-type Khokhlov–Semenov theory for thermotropics of semi-flexible polymeric nematogens must obey a self-consistent field equation that can be written as

$$\frac{1}{\sqrt{P(x)}} \nabla_x^2 \sqrt{P(x)} = -\lambda - \frac{2L_P \beta \Delta \epsilon}{L} S P_2(x), \qquad (6.6)$$

where β is the reciprocal thermal energy, $\Delta \epsilon$ the strength of van der Waals interaction that a chain experiences from all other chains, $P_2(x) = (3x^2 - 1)/2$ is the second Legendre polynomial, and $S = \langle P_2(x) \rangle$ the familiar nematic order parameter.

(b) In contrast to the original Maier–Saupe theory, it is now not so trivial to write down a self-consistent field equation for the nematic order parameter from Eq. (6.6). The reason is that the distribution function does not obey a simple Boltzmann distribution. Show that the trivial solution to the integro-differential equation (6.6) is the isotropic distribution. What value does the Lagrange multiplier take?

(c) Without actually solving Eq. (6.6), it is relatively straightforward to find the isotropic spinodal by using the finding of exercise 1 and using Eq. (1.10). Show that at the isotropic spinodal, we must have

$$\beta \Delta \epsilon |_{\text{spinodal}} = \frac{15}{2} \frac{L}{L_P}. \qquad (6.7)$$

(d) Argue why our finding that $\Delta \epsilon \propto L$ at the isotropic spinodal makes physical sense.

(e) Since the clearing temperature T_{IN} is close to the isotropic spinodal temperature, we are led to conclude that the nematic transition seems to move to higher temperatures the larger the persistence length L_P is for chains with fixed contour length L. This implies that the nematic phase becomes more stable. Explain why this is to be expected.

(3) **McMillan-type theory for hard-rod smectics.** In exercise 3 of Chap. 6, we worked out a simplified version of the McMillan theory for the nematic–smectic A transition in thermotropic systems in the spirit of the Maier–Saupe theory for the isotropic–nematic transition. Hard rods are known to also exhibit a nematic–smectic A transition, provided the aspect ratio is larger than some smallish critical value [29]. Interestingly, the transition is only weakly dependent on the aspect ratio of the particles, implying that the relevant concentration scale is the volume fraction of the particles ϕ. The transition occurs for a volume fraction of about 0.46 for very long rods.

(a) Describe how we can modify McMillan theory of exercise 3 of Chap. 6 for it to describe the nematic–smectic A transition in dispersions of hard rods.

(b) Why do we need not worry about equal pressures and chemical potentials to find the transition from the nematic to the smectic phase for perfectly parallel hard rods?

References

1. A. R. Khokhlov and A. N. Semenov, *On the theory of liquid-crystalline ordering of polymer chains with limited flexibility*, J. statist. Phys. **38** (1985), 161.
2. A. R. Khokhlov and A. N. Semenov, *Liquid crystalline ordering in the solution of long persistent chains*, Physica A **108** (1981), 546.
3. A. R. Khokhlov and A. N. Semenov, *Liquid crystalline ordering in the solution of partially flexible macromolecules*, Physica A **112** (1982), 605.
4. A. N. Semenov and A. R. Khokhlov, *Statistical physics of liquid-crystalline polymers*, Sov. Phys. Usp. **31** (1988), 988.
5. R. Colby and M. Rubinstein, *Polymer physics* (OUP, Oxford, 2004).
6. G. J. Vroege and H. N. W. Lekkerkerker, *Phase transitions in lyotropic colloidal and polymer liquid crystals*, Rep. Prog. Phys. **55** (1992), 1241.
7. Z. Y. Chen, *Nematic ordering in semiflexible polymer chains*, Macromolecules **26** (1993), 3419.
8. T. Odijk, *Theory of lyotropic polymer liquid crystals*, Macromolecules **19** (1986), 2313.
9. M. A. Cotter, *Generalized van der Waals theory of nematic liquid crystals: an alternative formulation*, J. Chem. Phys. **66** (1977), 4710.
10. W. M. Gelbart and B. Barboy, *van der Waals Picture of the Isotropic-Nematic Liquid Crystal Phase Transition*, Acc. Chem. Res. **13** (1980), 290.
11. A. Yu. Grosberg and A. R. Khokhlov, *Statistical physics of macromolecules* (AIP Press, Woodbury, NY, USA, 1994).
12. S. J. Picken, *Orientational order in nematic polymers - some variations on the Maier-Saupe theme*, Liquid Crystals, **37** (2010), 977.
13. Y. Jiang, C. Greco, K. Ch. Daoulas and J. Z. Y. Chen, *Thermodynamics of a compressible Maier-Saupe model based on the self-consistent field theory of wormlike polymer* , Polymers **9** (2017), 48.
14. P. van der Schoot and T. Odijk, *Statistical theory and structure factor of a semidilute solution of rodlike macromolecules interacting by van der Waals forces*, J. Chem. Phys. **97** (1992), 515.
15. C. Hiergeist, M. Lassig and R. Liposwky, *Bundles of interacting strings in 2 dimensions*, Europhys. Lett. **28** (1994), 103.
16. A. Speranza and P. Sollich, *Simplified Onsager theory for isotropic-nematic phase equilibria of length polydisperse hard rods*, J. Chem. Phys. **117** (2002), 5421.
17. W. Maier and A. Saupe, *Eine einfache molekular-statistische Theorie der nematischen kristallinflüssigen Phase. Teil I*, Zeitschrift für Naturforschung A, **14** (1960), 882.
18. W. Maier and A. Saupe, *Eine einfache molekular-statistische Theorie der nematischen kristallinflüssigen Phase. Teil II*, Zeitschrift für Naturforschung A, **15** (1960), 287.
19. L. Wu, E. A. Müller and G. Jackson, *Understanding and describing the liquid-crystalline states of polypeptide solutions: a coarse-grained model of PBLG in DMF*, Macromolecules **47** (2014), 1482.
20. M. J. Green, A. Nicholas G. Parra-Vasquez, N. Behabtu and M. Pasquali, *Modeling the phase behavior of polydisperse rigid rods with attractive interactions with applications to single-walled carbon nanotubes in superacids*, J. Chem. Phys. **131** (2009), 084901.
21. S. Chandrasekhar, *Liquid crystals*, 2nd edition (CUP, Cambridge, 1992).
22. S. Singh, *Phase transitions in liquid crystals*, Phys. Rep. **324** (2000), 107.

23. G. Vertogen and W. H. de Jeu, *Thermotropic liquid crystals, fundamentals* (Springer, Berlin, 1988).

24. Z. Dogic and S. Fraden, *Phase behavior of rod-like viruses and virus/sphere mixtures*, in: Soft Matter, Vol. 2, Complex Colloidal Suspensions, Eds G. Gompper and M. Schick (Wiley-VCH Verlag GmbH, Weinheim, 2006).

25. M. P. Allen, G. T. Evans, D. Frenkel and B. M. Mulder, *Hard convex body fluids*, Advances in chemical physics **86** (1993), 1–166.

26. H. N. W. Lekkerkerker and R. Tuinier, *Colloids and the depletion interaction* (Springer, Dordrecht 2011).

27. Z. Dogic, P. Sharma, M. Zakhary, *Hypercomplex liquid crystals*, Ann. Rev. Condens. Matter Phys. **5** (2014), 137–157.

28. M. Stone and P. Goldbart, *Mathematics for physics* (CUP, Cambridge, 2000).

29. P. Bolhuis and D. Frenkel, *Tracing the phase boundaries of hard spherocylinders*, J. Chem. Phys. **106** (1997), 666.

Chapter 7
Glossary

Anisometric   Not the same dimensions in all directions.

Anisotropic   Opposite of isotropic. Different in different directions.

Ansatz   An educated guess about the form of a function or an assumption about an outcome so as to make headway in solving an equation or cracking a problem.

Aspect ratio   The ratio between the length and breadth of a rigid body.

Athermal   A thermodynamic state weakly dependent on the temperature, typically dominated by entropy.

Biaxial   A quantity characterised by two symmetry axes.

Biaxial nematic   A nematic liquid crystal with two main symmetry axes and as a result three optical axes.

Bifurcation analysis   Mathematical analysis of under what conditions additional solutions to some non-linear equation or a set of equations branch off from a known solution.

Binodal   A collection of state points representing different states of matter that co-exist and that are in thermodynamic equilibrium with each other.

Body-axis vector   Unit vector representing the main axis of a rigid body, typically the axis associated with the largest dimension.

Boiling   Process of the transition of a liquid into a gas.

Boltzmann distribution   Equilibrium distribution describing a collection of particles, proportional to an exponential function of the negative of the ratio of a free energy and the thermal energy. Valid in the classical limit.

Boltzmann statistics   See classical statistics.

Born radius   Essentially the size of an atom. At centre-to-centre distances smaller than the sum of the Born radii of two atoms, their interaction becomes very strongly repulsive.

Calamitic nematic   Nematic liquid crystal consisting of rod-like particles.

Chemical potential   Intensive thermodynamic state variable of a compound conjugate to the number of particles of this compound present in a system.

P. van der Schoot, *Molecular Theory of Nematic (and Other) Liquid Crystals*, SpringerBriefs in Physics, https://doi.org/10.1007/978-3-030-99862-2_7

Chiral   Property of a body with a mirror image that cannot be superimposed on itself by a simple translation or rotation.

Cholesteric phase   Twisted, chiral nematic phase. The director field of a chiral nematic rotates in the direction perpendicular to it.

Chromonics   Dye molecules consisting of an aromatic core and polar side groups that self-assemble into supra-molecular structures that are highly anisometric in shape.

Classical statistics   Statistics of collections of particles in the classical limit of quantum statistics. Also known as Boltzmann statistics. For atoms and molecules, valid for temperatures above a few Kelvin.

Clearing temperature   The transition temperature between the isotropic and nematic phases.

Closed system   A thermodynamic system that exchanges heat with the environment but cannot do any mechanical work on it nor exchange particles with it.

Colloid   A material finely dispersed in another material in which it cannot be molecularly dissolved.

Colloidal particle   A particle in the size range from a few nanometres to a few microns, typically dispersed in a fluid.

Common tangent construction   Method to find states in thermodynamic equilibrium with each other (binodals). Makes use of a free energy per unit volume, expressed in terms of the particle density or concentration for a fixed temperature. It ascertains equality of chemical potentials, pressures, and temperatures in the co-existing phases.

Complex fluids   A collective noun for typically (but not exclusively) multi-component fluids, characterised by internal structures on length scales from several nanometres to several micrometres. Also known as structured fluids or soft matter. Includes liquids, solutions, colloids, polymers, foams, gels, glasses, liquid crystals, and many materials of biological origin. Although strictly not fluids, granular materials and plastic crystals are often also seen as representatives of complex fluids.

Condensation   Transition from a gas or dilute solution (or dispersion) to a liquid or dense solution (or dispersion).

Configuration   An arrangement of particles in space in which their positions and orientations are fixed.

Continuous phase transition   Also referred to as a second order phase transition, where the relevant order parameter changes continuously across the transition from one state of aggregation of a material to another. Certain thermodynamic susceptibilities diverge at the continuous transition point.

Correlation   Statistical dependence of two quantities.

Correlation function   Function that describes the covariance of two stochastic quantities.

Correlation hole   A volume in space near a test particle where another particle has a low (or zero) probability to be also present.

Critical point   A thermodynamic state separating two distinct states of matter characterised by a gradual transition.

Critical exponent Near a critical point, the values of thermodynamic properties of a material tend to depend on the distance to that critical point raised to some power. The power is called the critical exponent. The distance to the critical point is often measured in terms of the difference between the temperature and the critical temperature.

Degrees of freedom For classical particles, their positions and orientations in space.

Density functional theory In classical systems, a theory in which the appropriate free energy is written as a functional of the density distribution in position or orientation space.

Director Average orientation of particles in a liquid crystal. Typically averaged over a mesoscopic volume.

Discontinuous phase transition Also referred to as a first order phase transition, where the relevant order parameter changes discontinuously across the transition from one state of aggregation of a material to another. Often (but not always) associated with a latent heat.

Discotic nematic Nematic liquid crystal consisting of disk-like particles.

Divergence Attaining a positive or negative infinite value.

Diverge The process of a quantity attaining a positive or negative infinite value.

Enthalpy Thermodynamic potential (or state function) associated with an isothermal–isobaric system, given by the sum of the internal energy and the product of the pressure and volume.

Entropy Thermodynamic state function that measures the degree of disorder in or uncertainty about the microscopic state of a system.

Equal area construction Method to find states in thermodynamic equilibrium with each other (binodals). It makes use of the equation of state, describing the pressure as a function of the volume of a given amount of materials at a fixed temperature. Enforces equality of chemical potentials, pressures, and temperatures in the co-existing phases. Also known as Maxwell construction.

Equation of state Relation between thermodynamic variables, such as pressure, density, and temperature, for phases of matter in a state of equilibrium.

Excluded volume Volume that one particle excludes another due to the short-range repulsion acting between them. Also known as co-volume.

Deflection length Length scale over which bending fluctuations of a semi-flexible polymer are suppressed due to the presence of an external or molecular ordering field. Also known as the Odijk length.

Free energy A thermodynamic state function describing the amount of energy of a system that can be converted into useful work, depending on how that system interacts with its environment.

Free volume Volume of a material not taken up by particles. See also excluded volume.

Freezing (transition) Solidification of a liquid.

Frozen-in degrees of freedom Degrees of freedom of a particle that do not attain all physically allowed values with the equal probability due to the presence of an external field or a molecular field caused by interactions with other particles.

Gel   Soft solid consisting of a network of colloidal particles or polymers in a fluid.

Half-open system   A thermodynamic system in which no mechanical work can be done on the environment, yet energy can be exchanged with that environment, as well as some (but not all) of the types of particles present in it.

Hard-core potential   Harshly repulsive model potential between particles, described by an infinite energy of overlapping particles and zero energy for non-overlapping particles.

Hard particles   Hypothetical particles that interact solely via a hard-core potential.

Ideal gas   A gas obeying the ideal gas law. Sufficiently dilute gases at temperatures well above absolute zero behave according to the ideal gas law.

Ideal gas law   Equation of state of a gas in which the pressure is equal to the number density of the gas times the thermal energy of the gas.

I-N transition   Transition between the isotropic and nematic phases of a material.

Invariant   A property of a quantity that remains unchanged under some transformation or in response to a change in value of another quantity.

Isolated system   A system that does not exchange energy and particles with the environment and cannot do mechanical work on it.

Isotherm   Values that a thermodynamic state variable takes at a constant temperature.

Isotropic   Equal in all directions of space.

Lagrange multipliers   Parameters that enforce external conditions on a set of variables, when maximising or minimising a function of those variables.

Lag time   Time it takes for the response of a material to take off following a sudden change in the thermodynamic conditions. Typically associated with discontinuous transitions between states of matter.

Lattice statistics   Statistical mechanics of particles in which continuous space is replaced by a lattice, where the particles are forced to occupy cells associated with the lattice points.

Law of corresponding states   An equation of state that, if expressed in appropriately reduced units, is valid for a large class of materials.

Legendre polynomials   A particular kind of a complete set of orthogonal polynomials.

Lennard–Jones potential   Model atomic potential, characterised by a short-range repulsion and a long-range attraction. Also known as a "6 − 12" potential due to the negative sixth and negative twelfth power of the distance dependence of the attraction and the repulsion, respectively.

Lennard–Jonesium   Hypothetical material consisting of particles that interact via the Lennard–Jones potential.

Lyotropic liquid crystal   Liquid crystalline state of particles dispersed in a fluid medium. The relevant control variable is the concentration of the particles. In a more restrictive sense, liquid crystals in solutions of amphipolar molecules known as surfactants.

Marginal thermodynamic stability   Boundary of thermodynamically stable uniform phases. See also spinodal.

Maxwell construction   (See) Equal area construction.

Mean-field theory   A theory in which the interaction of a particle with all other particles is replaced by some average (mean) field.

Mesogen   A molecule or particle that under appropriate conditions forms aggregated states in between the common disordered liquid and ordered crystalline states.

Mesophase   Aggregated state characterised by partial long-range order in orientation or position.

Metastable phase   State that represents a local free energy minimum but not a global free energy minimum.

Microstate   Particular realisation or conformation of a collection of particles, in which their positions and orientations in space are given.

Molecular field   The interaction free energy a particle experiences from the presence of all other particles in the system. Acts, in a way, like an external field.

Monodisperse particles   Particles of (ideally) the same dimensions and chemical composition.

Nematogen   A molecule or particle that under appropriate conditions forms a nematic liquid crystal.

Non-ideal gas   Opposite of an ideal gas, in which the particles interact sufficiently strongly for the pressure not to obey the ideal gas law.

Nucleation and growth   Phase separation process of a material suddenly brought in a metastable thermodynamic state. It progresses via the nucleation of a bubble, drop, or crystal of the stable state and subsequent growth of that state as a function of time.

Odijk length   See deflection length.

Order parameter   A quantity that describes the degree of order of a material system. If expressed in the form of a scalar, usually defined to be equal to zero if that system is in its disordered state, and unity if the order is perfect (maximal).

Orientational distribution function   Function describing the probability distribution of the orientation particles in space.

Osmotic pressure   Excess pressure of a solution or dispersion relative that of the pure solvent it is in thermal and partial chemical equilibrium with. Partial equilibrium is established via a semi-permeable membrane that allows solvent but no solute transport across it.

Pair correlation function   A function that provides a measure for the probability that at a distance from a given test particle another particle can be found.

Pair potential   Interaction potential between two particles.

Persistence length   Length scale along the backbone of a polymer over which the polymer molecule (on average) loses its sense of direction due to thermal agitation.

Phase gap   Difference in concentrations or density of co-existing phases.

Phase diagram   Diagram showing what phases of a material are stable under what thermodynamic conditions, typically as a function of temperature and pressure or density.

Phase transition   Transition from one state of aggregation to another.

Phenomenological   Based on experience and intuition, not derived from first principles or fundamental laws.

Plastic crystal   Crystal phase of particles of which one or more orientational degrees of freedom are not frozen-in.

Polydisperse particles   Particles of different sizes, shapes, and/or chemical composition.

Polymer   Chain-like molecule formed by chemically linking large numbers of one or more types of molecular subunit.

Potential of mean force   An average potential between particles in which certain molecular details, degrees of freedom, or properties are ignored and described in an average fashion. Potentials of mean force are free energies rather than bare energies.

Quench   A sudden (instantaneous) change of thermodynamic conditions.

Reduced units   Units of physical quantities that are made dimensionless by dividing them or expressing them in terms of (groups of) appropriate fundamental physical quantities.

Scalar nematic order parameter   A number between negative one-half and unity describing the degree of alignment of nematogens along the director, with zero representing the disordered state and unity the perfectly ordered state. Negative values describe a state where the particles on align to some degree at right angles to the director.

Self-assembly   Spontaneous and reversible assembly of particles to form larger aggregates that themselves may be viewed as particles.

Self-consistent field theory   A theory in which the equation of state describing a physical property is a function that has same property, such as in a Boltzmann distribution of a probability that depends on that same Boltzmann probability. It describes the response of a particle to a molecular field that is a function of the properties of that same particle.

Semi-flexible polymer   Polymer molecule that is locally rigid and has a persistence length that is very much larger than the width of the molecule or the size of the molecular subunit. The contour length of the chain is tacitly presumed to be much larger than the persistence length; otherwise, it would be called semi-rigid.

Soft-core interaction   An interaction for which the pair potential changes gradually with distance between particles and is strongest for short separations.

Solvation shells or layers   Accumulation of particles in shells or layers around a test particle due to attractive interactions between them and/or crowding at high densities.

Soft (condensed) matter   Condensed matter, typically but not exclusively involving a fluid component, characterised by internal structures on length scales from several nanometres to several micrometres. Includes liquids near interfaces, solutions, colloids, polymers, foams, gels, glasses, liquid crystals, plastic crystals, and many biological materials. Also known as complex or structured fluids.

Special functions   Widely used functions that have a name attached to them, sometimes without a known explicit form and defined by a differential equation or an integral.

Spherical harmonics   A complete set of normalised and orthogonal functions forming a so-called orthonormal basis.

Spinodal   Limit of thermodynamic stability.

Spinodal decomposition   Dynamics of a spontaneous phase transition taking place in a thermodynamically unstable state following a quench.

Statistical moment   Expectation value of some power of a stochastic variable. For the first moment of a distribution function, the power is unity, for the second moment two, and so on.

Symmetry   Property defined by an invariance under particular kinds of transformation, such as translation, rotation, or reflection. In the context of crystals, it refers to the set of such operations on the crystal lattice that leaves it unaltered.

Symmetry breaking   Transition in which the symmetry of a material changes.

Sublimation   Formation of a solid out of a gas.

Super critical fluid   Fluid state above the critical temperature, where the distinction between a gas and a liquid ceases to exist.

Thermal energy   The product of Boltzmann's constant and the absolute temperature. The relevant energy scale of classical (Boltzmann) statistics.

Thermotropic liquid crystal   Liquid crystal formed by molecules or polymers. The relevant control parameter is the temperature.

Tie line   Line connecting the densities or concentrations between co-existing phases.

Trial function   See ansatz.

Triple point   Thermodynamic state point in which three states of matter are found to co-exist, typically (but not exclusively) a gas, a liquid, and a solid.

Universal   Valid for a very large group of materials of a certain kind, irrespective of their chemical makeup.

Universality (class)   The principle that certain behaviour is equivalent for (groups of) materials of a certain kind.

van der Waals equation of state   An equation of state of a non-ideal (interacting) gas due to Johannes Diderik van der Waals.

van 't Hoff's law   Equivalent of the ideal gas law for the osmotic pressure of dilute solutions or dispersions.

Virial co-efficient   A co-efficient of the virial expansion of the pressure of a gas or osmotic pressure of a dispersion.

Virial expansion   Expansion of the pressure or osmotic pressure in terms of a sum of the product of co-efficients and powers of the density.

Chapter 8
Solutions to the Exercises

Here, we give brief solutions to the exercises provided at the end of each chapter.

1. Introduction

(1) **Van der Waals and Lennard–Jones connected**

  (a) Taylor expanding the van der Waals equation of state to linear order in the small quantity $b\rho \ll 1$ gives $p = \rho k_B T (1 - b\rho)^{-1} - a\rho^2 = \rho k_B T (1 + b\rho + \cdots) - a\rho^2$. Rearranging and expressing the pressure as the ideal gas law times a correction, gives $p = \rho k_B T \times [1 + \rho(b - \beta a) + \cdots]$ with $\beta \equiv 1/k_B T$. Hence, the second virial co-efficient reads $B_{vdW} = b - \beta a$.

  (b) Use the expression $B_{LJ} = 2\pi \int_0^\infty dr r^2 [1 - \exp(-\beta U_{LJ})]$ and insert the expression for the Lennard–Jones potential $U_{LJ}(r) = 4\epsilon \left[t^{-12} - t^{-6}\right]$ with $t \equiv r/\sigma$. This gives, invoking the co-ordinate transformation $r = \sigma t$, $B = 2\pi\sigma^3 \int_0^\infty dt\, t^2 \left[1 - \exp\left[-4\beta\epsilon(t^{-12} - t^{-6})\right]\right]$. The exponent is essentially nil for $t \leq 1$ and to leading order in $t > 1$ equal to $1 - 4\beta\epsilon t^{-6}$. Carrying out the integrations gives $B = 2\pi\sigma^3/3 - 8\pi\sigma^3\beta\epsilon/5$.

  (c) Demanding that $B_{LJ} = B_{vdW}$, we conclude from (a) and (b) that $b = 2\pi\sigma^3/3$ and $a = 8\pi\epsilon/5$.

  (d) For high temperatures, $\beta\epsilon \to 0$. That the integral is not zero even if $\beta\epsilon$ is very small is because the integral is then dominated by the contribution from the power t^{-12} in the integral in the domain from 0 to 1. This is the term due to the strong repulsion of particles at short separations. Hence, $\lim_{\beta\epsilon \to 0} B_{LJ} = 2\pi\sigma^3 \int_0^1 dt\, t^2 [1 - \exp(-4\beta\epsilon t^{-12})] \sim 2\pi\sigma^3 \int_0^{(4\beta\epsilon)^{1/12}} dt\, t^2 \sim \frac{2}{3}\pi\sigma^3(4\beta\epsilon)^{1/4}$. Here, we made use of the co-ordinate transformation $s = t/(4\beta\epsilon)^{-12}$ and the fact that the exponent drops to zero very quickly

© The Author(s), under exclusive license to Springer Nature Switzerland AG 2022
P. van der Schoot, *Molecular Theory of Nematic (and Other) Liquid Crystals*,
SpringerBriefs in Physics, https://doi.org/10.1007/978-3-030-99862-2_8

for $t > (4\beta\epsilon)^{12}$. Hence, we find that $B_{LJ} \propto \sigma^3(\beta\epsilon)^{1/4}$ approaches zero in the limit $\beta\epsilon \to 0$, but only very slowly so due to the one-fourth power.

(2) **Thermodynamics and the van der Waals equation of state.**

(a) The isothermal compressibility β (not to be confused with the reciprocal thermal energy!) is defined as $\beta = -V^{-1}(\partial V/\partial P)_{N,T}$. Converting the volume V in this expression into a density $\rho = N/V$ gives $\beta = \rho^{-1}\partial\rho/\partial p|_{N,T}$. Since $\beta > 0$, we have $\partial p/\partial\rho \geq 0$. Furthermore, from thermodynamics we have $p = -\partial F/\partial V|_{N,T} = -\partial(Vf)/\partial V|_{N,T} = -f + \rho\partial f/\partial\rho|_{N,T}$. Hence, $\partial f/\partial\rho = \rho\partial^2 f/\partial\rho^2 \geq 0$, implying that $\partial^2 f/\partial\rho^2 \geq 0$.

(b) Rewrite the van der Waals equation of state to $p = \rho k_B T(1 - \rho b)^{-1} - a\rho^2$. Using this expression and demanding that in stable thermodynamic equilibrium $\partial p/\partial\rho \geq 0$ imply that this is the case provided $k_B T \geq 2a\rho(1 - b\rho)^2$. So, for sufficiently low temperatures this bound is violated and the homogeneous fluid cannot be stable.

(c) The spinodal is given by the condition $\partial p/\partial\rho = k_B T(1 - b\rho)^{-2} - 2a\rho = 0$. The maximum in the spinodal temperature as a function of the density is found by demanding that $\partial^2 p/\partial\rho^2 = 0$. These two equations fix $\rho = \rho_c$ and $T = T_c$ at the critical point. The last equality gives $a = bk_B T_c(1 - b\rho_c)^{-3}$. Inserting this in the expression for the spinodal gives $b = 1/3\rho_c$. Combining these results leads to $a = 9k_B T_c/\rho_c$. Note that this fixes $\rho_c = 1/3b$ and $T_c = 8a/27bk_B$ for the critical density and temperature in terms of the van der Waals parameters.

(d) By inserting the expressions for a and b that we found in (c) in the van der Waals equation of state, we find for the critical pressure $p_c = 3\rho_c T_c/8$.

(3) **Van der Waals theory of freezing.**

(a) Carrying out the differentiation and making use of the chain rule, we obtain $\partial(Nf(\rho))/\partial V = N(\partial f(\rho)/\partial\rho)(\partial\rho/\partial V) = -\rho^2(\partial f(\rho)/\partial\rho)$, with $f \equiv F/Nk_B T$ representing the expressions in the square brackets in Eqs. (1.11) and (1.12) a function of ρ, produces the quoted equations of state. The alternative route requires integrating the equations of state as $F = -\int dVp|_{N,T}$. For instance, using for simplicity the ideal gas law $p = \rho k_b T = Nk_b T/V$, we obtain $F = -Nk_b T \ln V + C(N,T)$, with $C(N,T)$ an integration constant (in V). Note that $N \ln V$ is "super extensive" and does not scale linearly with system size due to the logarithm. To render the Helmholtz free energy extensive, we use $C(N,T)$ to make the logarithm intensive, and writing $C(N,T) = N \ln[Nw(T)/e]$ does that. Note that the constant ω cannot depend on N nor ρ because it needs to be intensive and is a constant of V. The same principle applies to the van der Waals equations of state.

(b) The internal energy contribution U to the Helmholtz free energy for hard particles has its source in the kinetic energy of the particles. This is also true for ideal gases, and hence, one needs to only focus on the ideal gas contribution. Using $U = \partial \beta F/\partial \beta$, where $\beta = 1/k_B T$, then gives for both the fluid and solid phases $U = N\partial \ln \omega/\partial \beta = 3N/2\beta$, or, $2d \ln \omega = 3d \ln \beta$ and $\omega \propto \beta^{3/2}$. In conclusion, we surmise that $\omega \propto T^{-3/2}$.

(c) We can define a shifted Helmholtz free energy by subtracting $N \ln \omega/b$ from Eqs. (1.11) and (1.12), which eliminates the ω. The shift does not affect the pressure and only shifts by the same amount the chemical potential of the particles in the fluid and solid phases, so cannot affect the co-existence of the fluid and solid phases.

(d) The chemical potential and pressure follow from $\mu = \partial F/\partial N|_{V,T} = \partial f/\partial \rho$, where we now define $f \equiv F/V$, and recall from exercise (2) that $p = -f + \rho \partial f/\partial \rho|_{N,T}$. Hence, $f = -p + \rho \partial f/\partial \rho|_{N,T} = -p + \rho \mu$. The common tangent to the function $f(\rho)$ produces straight line that touches this function from below. The points that the line touch share the same slope and hence chemical potential and must also have the same intercept and hence pressure. See Fig. 8.1.

(e) Set $\rho b = \phi/0.64$ in Eq. (1.11) and $\rho \hat{b} = \phi/0.74$ in Eq. (1.12), and plot $\beta f = \beta F/V$ as a function of ϕ for the two phases to identify, e.g., graphically, the common tangent.

(f) The free energy densities βf for the two phases cross just below $\phi = 0.64$, implying that the free energy of the fluid state becomes larger than that of the solid at equal density. Since the model has no potential energy and the temperatures in the co-existing phases are equal, the kinetic energies per particle must be equal. This implies that the internal energy U remains fixed, and only entropy contributes to any changes in the free energy. The entropy of the (stable) crystal phase must therefore be larger than that of the (unstable) fluid phase, as $F = U - TS$.

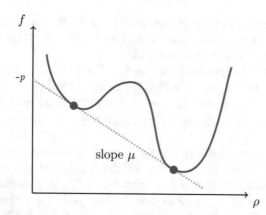

Fig. 8.1 Common tangent construction. Plotted is the free energy per unit volume, f, versus the number density of particles, ρ, for a given fixed temperature. The free energy density, represented by the drawn line, has two minima. The slope of this curve represents the chemical potential, μ. The straight line (dashed) that touches the two minima (indicated by the dots) identifies two phases in thermodynamic with equal temperature, chemical potential, and pressure. The abscissa of the straight line is the negative pressure, $-p$.

2. Liquid Crystals

(1) Cooling down thermotropics.

(a) The amount of energy $H \geq 0$ extracted from the material obeys $H = -C_p \Delta T$, where $\Delta T \leq 0$ [T] its change in temperature. Since we also have $H = Rt$, we conclude that $\Delta T = -H/C_p = -Rt/C_p$ drops linearly with time, except near the clearing temperature T_{IN}. At the clearing temperature for all intents and purposes $\Delta T = 0$, as C_p diverges at that point. The temperature remains a constant $T = T_{IN}$ until all of the latent heat H_{IN} is extracted. This takes H_{IN}/R seconds. After that has happened, all of the isotropic phase has transitioned and C_p becomes finite again but has a different value from that in the isotropic phase. The temperature then starts to decrease linearly again in proportion with the amount of energy extracted.

(b) The answer is implicit in (a): the lowering of the temperature stalls at a $T = T_{IN}$ until the latent heat is extracted.

(c) Let V_N be that part of the volume V of the fluid that is in the nematic state. For $T > T_{IN}$, we have $V_N/V = 0$. For $T < T_{IN}$, we have $V_N/V = 1$. Since the latent heat H_{IN} is an extensive quantity, the volume of nematic phase grows linearly with the amount of energy extracted. So, as a function of H, $V_N/V = 0$ until we hit the clearing temperature, after which V_N/V grows linearly with H, until all of the isotropic phase is converted, following which $V_N/V = 1$.

(2) Distribution functions and order parameters.

(a) The first moment $\langle \vec{n} \cdot \vec{u} \rangle$ is zero on account of inversion symmetry: $P(\vec{u}) = P(-\vec{u})$.

(b) A dipole aligns with an electric field. If the dipole moment is at some angle to the main body axis vector of the particle, the main body axis vector will align with some angle to that field.

(c) For perfectly aligned rods, we have $\vec{n} \cdot \vec{u} = 1$ so $S = \langle (3-1)/2 \rangle = 1$ due to the normalisation of the distribution function. If perfectly "disordered", $\vec{n} \cdot \vec{u} \equiv 0$, and $S = \langle -1/2 \rangle = -1/2$. For randomly oriented rods, $\langle (\vec{n} \cdot \vec{u})^2 \rangle = 2^{-1} \int_{-1}^{+1} dx x^2 = 2/3$, where $x = \cos \theta$ and $P = 1/4\pi$. So, $S = 0$ in that case. See Eq. (2.2).

(d) The order parameter should be invariant to inversion symmetry, so one expects S to be a function of $(\vec{n} \cdot \vec{n})^2$ that equals unity for perfect alignment and zero for an isotropic distribution. In 2D, one would choose $S = \langle 2(\vec{n} \cdot \vec{u})^2 - 1 \rangle = \langle \cos 2\theta \rangle$, where $\theta = \arccos(\vec{n} \cdot \vec{u})$. For perfect order, this produces $S = 1$, and for isotropic orientations, $S = 0$. For a more formal derivation of the scalar order parameter from a tensorial order parameter can be found in, e.g., the didactic review by Denis Adrienko [1].

(3) **Van der Waals picture of nematics.**

(a) Since a and b are presumed to not depend on the temperature, we must conclude that for the entropy $S = Nk_B \ln(1 - \rho b)$. Since $1 - \rho b < 1$, we must have $\ln(1 - \rho b) < 0$, and for that reason $S < 0$. Notice that this is an excess entropy relative to that of the ideal gas. Packing in more particles increases the entropy loss with increasing density ρ. The reason can only be a reduction of free volume with increasing particle density.

(b) The excluded volume of two parallel elongated particles is smaller than that of two elongated particles at right angles. This is why more matches fit in a matchbox if put in perfectly parallel and in register than if one attempts to insert a pile of randomly oriented matches into that same matchbox. Hence, b decreases with increasing value of the nematic order parameter. The van der Waals attraction decreases with distance of two atoms. The mean distance between pairs of atoms, part of two elongated molecules, increases with increasing angle between these molecules. Hence, the strength of the van der Waals attraction increases with decreasing angle between them.

(c) Because any type of alignment decreases the excluded volume of elongated particles, which includes not only alignment that leads to a positive order parameter, but also causes the order parameter to be negative. Hence, b should decrease with some even power of S. Since the van der Waals attraction increases with increasing degree of order, positive or negative, a should also increase with an even power of the order parameter. In terms of a series expansion in even powers of S, the first correction term should scale as S^2.

(d) See (a) and (b): this can only be translational entropy as it is associated with free volume, i.e., the volume available to the centres of mass of the particles for translational motion.

(e) We expect the orientational entropy to decrease with increasing degree of orientational order. In a description where the parameters a and b obtain corrections of order S^2, we could add an orientational entropy loss term to the free energy, such as $c_2 Nk_B T S^2 + c_3 Nk_B T S^3 + \cdots$. The proportionality to T shows it is the correction of entropic character. Further, it is proportional to the number of particles, as it should, and any kind of ordering goes at the cost of orientational entropy, so the leading order term should at least be quadratic in the order parameter. We expect $c_2 > 0$, but c_3 could be positive or negative.

3. The Groundwork

(1) **Ideal gases of point particles revisited.**

   (a) If $\beta = 1/k_B T$, we have $\beta F_{id} = N \ln \rho \Lambda^2/e$. According to thermodynamics, $\beta p = -\partial \beta F/\partial V|_{N,T}$. Inserting $F = F_{id}$ gives $\beta p = \rho$. This is the ideal gas law.

   (b) From thermodynamics, we have $\mu = \partial F/\partial N|_{V,T}$. For $F = F_{id}$, this becomes $\beta \mu = \ln \rho \Lambda^3$.

   (c) From (a) and (b), we deduce that $\partial \beta \mu/\partial V|_{N,T} = -1/V$ for an ideal gas, and $\partial \beta p/\partial N|_{V,T} = 1/V$, so $\partial \beta \mu/\partial V|_{N,T} = -\partial \beta p/\partial N|_{V,T}$ in agreement with the quoted Maxwell relation.

   (d) Write $\partial(F/T)/\partial(1/T) = \partial(F \times 1/T)/\partial(1/T)$, and apply the chain rule, $\partial(F \times 1/T)/\partial(1/T) = \partial F/\partial(1/T) \times (1/T) + F = -T^2(\partial F/\partial T) \times (1/T) + F = -T\partial F/\partial T + F$. Realise that $S = -\partial F/\partial T$, so that $\partial(F/T)/\partial(1/T) = TS + F = TS + (E - TS) = E$.

   (e) Use $E = \partial \beta F/\partial \beta|_{N,V}$, and insert $\beta F = \beta F_{id} = \ln \rho \Lambda^3$ with $\Lambda = \sqrt{\beta h^2/2\pi m}$, to obtain $E = N(\rho \Lambda^3)^{-1} \times \rho 3\Lambda^2 \times \partial \Lambda/\partial \beta|_{N,V}$. Further, $\partial \Lambda/\partial \beta|_{N,V} = \Lambda/2\beta$. Hence, $E = 3N/2\beta = 3Nk_B T/2$. This is of course equal to the mean kinetic energy of the particles in three-dimensional space, and we recognise the law of equipartition of energy in it.

   (f) For the isochoric heat capacity, we have the definition $C_V = \partial E/\partial T|_{N,V} = 3Nk_B$, which does not depend on temperature, as advertised.

(2) **Ideal polar particles in an external field.**

   (a) Straightforward integration gives $\int d\vec{u} P(\vec{u}) = 2\pi \int_{\cos \theta_0}^{+1} d \cos \theta \, [2\pi(1 - \cos \theta_0)]^{-1} = (1 - \cos \theta_0)^{-1}(1 - \cos \theta_0)] = 1$.

   (b) $\langle \cos \theta \rangle = 2\pi \int_{\cos \theta_0}^{1} d \cos \theta \cos \theta [2\pi(1 - \cos \theta_0)]^{-1} = (1 - \cos \theta_0)^{-1}$ $\int_{\cos \theta_0}^{+1} dx x = (1 - \cos \theta_0)^{-1}(1 - \cos \theta_0^2)/2$. Hence, $\langle \cos \theta \rangle = (1 + \cos \theta_0)/2$. Note that for $\theta_0 = \pi$, so zero field and random orientations, we have $\langle \cos \theta \rangle = 0$. For perfectly aligned particles, $\theta_0 \to 0$, and $\langle \cos \theta \rangle = 1$.

   (c) Writing out the square gives $\sigma_{\cos \theta}^2 = \langle (\cos \theta - \langle \cos \theta \rangle)^2 \rangle = \langle (\cos^2 \theta - 2\langle \cos \theta \rangle \cos \theta + \langle \cos \theta \rangle^2) \rangle$. Note that $\langle \cos \theta \rangle$ is a constant and can be taken out of the averages. Hence, $\sigma_{\cos \theta}^2 = \langle (\cos^2 \theta) - 2\langle \cos \theta \rangle \times \langle \cos \theta \rangle + \langle \cos \theta \rangle^2 \rangle$ can also be written as $\langle \cos^2 \theta \rangle - \langle \cos \theta \rangle^2$.

   (d) Simple integration gives $\langle \cos^2 \theta \rangle = (1 - \cos^3 \theta)/2(1 - \cos \theta)$. We have already seen that $\langle \cos \theta \rangle = (1 + \cos \theta_0)/2$. For $\theta_0 = \pi$, we find $\sigma_{\cos \theta}^2 = 1/3$ and $\langle \cos \theta \rangle = 0$, implying that the standard deviation $\sigma_{\cos \theta} = 1/\sqrt{3}$ is large in comparison with the expectation value. For $\theta_0 \to 0$, we make use of a Taylor expansion to quadratic order in $\theta_0 \ll 1$ and obtain $\sigma_{\cos \theta}^2 = \theta_0^2/2 + \cdots \ll 1$. As $\langle \cos \theta \rangle = 1$, we now have the opposite case, with the standard deviation small relative to the average.

(e) The Gibbs entropy follows from Eq. (3.4), realising that in this case $S = -F/N$, which in the present case simplifies to $S/Nk_B = -2\pi \int_{\cos\theta_0}^{+1} d\cos\theta \, [2\pi(1 - \cos\theta_0)]^{-1} \ln[2\pi(1 - \cos\theta_0)]^{-1} = \ln[2\pi(1 - \cos\theta_0)]$. Hence, for $\theta_0 = \pi$, $S/Nk_B = \ln 4\pi$. For $\theta_0 \ll 1$, we find by Taylor expansion $S/Nk_B = \ln\pi\theta_0^2$. The difference in entropy of the two states is $\Delta S/Nk_B = \ln\theta_0^2/4$, which becomes negative infinite if θ_0 approaches zero.

(3) Lennard–Jonesium.

(a) The Lennard–Jones potential can be made universal by realising that $U_{LJ}/\epsilon = 4[(r/\sigma)^{-12} - (r/\sigma)^{-6}]$. So, if we define $u_{LJ} \equiv U_{LJ}/\epsilon$ and $x \equiv r/\sigma$, we obtain $u = 4[x^{-12} - x^{-6}]$. The potential is obviously zero for $x = 1$. We find the minimum value of u by setting $du/dx = 4(-12x^{-13} + 6x^{-7}) = 24x^{-7}(-2x^{-6} + 1) = 0$. Hence, for $x = 2^{1/6} \approx 1.12$, we find the minimum value of the potential, equal to $u = -1$. For $x < 1$, the potential strongly rises with decreasing distance to diverge for $x = 0$, while for $x > 1$ it rises slowly from the minimum of $u = -1$ with increasing distance to a value $u = 0$ at infinite distance.

(b) The force F_r acting between two particles along the line connecting the centres of mass is equal to $F_r = -dU/dr$. Hence, for $x < 1$, the force is positive as the potential has a negative slope, while for $x > 1$ the slope of the potential is positive and hence the force negative. This means that for $x < 1$ the particles repel each other, and for $x > 1$ they attract each other. Recall that the potential has a minimum for $x = 1$.

(c) The curves look like a smooth version of the Heaviside step function, with $g(x) \approx 1$ for $x > 1$ and $g(x)$ for $0 \leq x \leq 1$, if $x = r/\sigma$. There is an overshoot with a maximum of $\exp\beta\epsilon$ at $x = 2^{1/6} \approx 1.12$. The larger the temperature, the smaller $\beta\epsilon$, and the closer the maximum value is to unity.

(d) The maximum in the pair correlation function represents the first so-called solvation shell of the particles: every particle is surrounded by a shell of other particles. The larger the temperature the less pronounced this shell is and hence the less structured the fluid is.

(e) Since $g \approx 0$ for $x < 1$, σ must represent the distance over which two particles cannot occupy the same space, i.e., the hard-core diameter of the particles. This is also called the Born diameter of the particles, i.e., the size of the particles.

(f) For $\beta\epsilon \ll 1$, the pair correlation function resembles a Heaviside step function. Within a Boltzmann description, this is well represented by the hard-core potential.

4. Onsager Theory

(1) **Dilute dispersion of dipolar rod-like particles in an external field.**

    (a) For the functional minimisation of the Gibbs entropy, follow the recipe as prescribed in Eqs. (4.10) and (4.11). For that of the external field, and the normalisation involving the Lagrange parameter λ, use $F_{ext}[P + \delta P] = \int d\vec{u}[P(\vec{u}) + \delta P(\vec{u})]U_{ext}(\vec{u}) = F_{ext}[P] + \int d\vec{u}\delta P(\vec{u})U_{ext}(\vec{u})$. Hence, $\delta F_{ext}/\delta P(\vec{u}) = U_{ext}(\vec{u})$. This gives for the equilibrium distribution $P(\vec{u}) = \exp[\lambda - 1 - \beta U_{ext}(\vec{u})]$. We fix $\exp(\lambda - 1)$ by insisting that $\int d\vec{u} P(\vec{u}) = 1$. This gives Eq. (4.14).

    (b) If $x \equiv \vec{u} \cdot \vec{n}$, $Z = \int d\vec{u} \exp(-\beta U_{ext}(\vec{u})) = \int_0^{2\pi} d\phi \int_{-1}^{+1} dx \exp(-\beta p E x) = 2\pi(\beta p E)^{-1} \int_{-\beta p E}^{+\beta p E} dt \exp(-t)$. If $K = \beta p E$, then simple integration gives $Z = 4\pi K^{-1} \sinh K$.

    (c) We have $\langle x \rangle = 2\pi \int_{-1}^{+1} dx x P(\vec{u}) = 2\pi Z^{-1} \int_{-1}^{+1} dx x \exp(Kx)$. Since $Z = 2\pi \int_{-1}^{+1} dx \exp(Kx)$, we have $dZ/dK = 2\pi \int_{-1}^{+1} dx \partial(\exp(Kx))/\partial K = 2\pi \times \int_{-1}^{+1} dx x \exp(Kx)$. This confirms Eq. (4.15) for $n = 1$. Repeated differentiation gives the result for $n > 1$.

    (d) From (c), we have $\langle x \rangle = K^{-1} + \cosh K/\sinh K$. The electrostatic free energy is given by $\beta F_{ext}/N = K\langle x \rangle$. The orientational entropy is given by the Gibbs formula $S_{or} = -k_B N \int d\vec{u} P(\vec{u}) \ln P(\vec{u})$ because the particles do not interact and the entropy is extensive. Furthermore, by inserting the expression for P in the logarithm, and making use of the properties of averages, we conclude that $\int d\vec{u} P(\vec{u}) \ln P(\vec{u}) = -\langle x \rangle K + \ln Z$. Hence, we obtain for the orientational free energy $\beta F_{or} = -T S_{or} = K\langle x \rangle - \ln Z$. The behaviour of these quantities can be assessed by taking the limits $K \ll 1$ and $K \gg 1$. For $K \ll 1$, we have by Taylor expansion $\sinh K \sim K + K^3/6 + O(K^5)$, $\cosh K \sim 1 + K^2/2 + O(K^4)$. So, $\langle x \rangle \sim K/2 + \cdots$, as one would expect for a linear response, implying that $\beta F_{ext}/N \sim -K^2/2$. $\beta F_{or}/N \sim -\ln 4\pi + K^2/3 + \cdots$ increases, so entropy decreases with increasing field strength, as one would expect.

    (e) From (d), we conclude that the free energy $F = F_{or} + F_{ext} = -Nk_B T \ln Z + K\langle x \rangle - K\langle x \rangle = -Nk_B T \ln Z$. The variance we can calculate from $\sigma_x^2 = d^2 \ln Z/dK^2 = d[d\ln Z/dK]/dK = d[Z^{-1}dZ/dK]/dK = -Z^{-2}(dZ/dK)^2 + Z^{-1}d^2Z/dL^2 = \langle x^2 \rangle - \langle x \rangle^2$, where we have used Eq. (4.15). Note also that $\sigma_x^2 = d[d\ln Z/dK]/dK = d\langle x \rangle dK$, linking the susceptibility of the particle to an external field to the variance of the conjugate variable. This is a static variant of the fluctuation dissipation theorem, linking susceptibilities to the magnitude of spontaneous fluctuations [2].

(f) By differentiation, we find for the variance $\sigma_x^2 = d\langle x\rangle/dK = K^{-2} - \sinh^{-2} K$, recalling that $\langle x\rangle = -K^{-1} + \cosh K/\sinh K$. In the limit $K \ll 1$, we have $\sigma_x^2 \sim 1/\sqrt{3}$ and $\langle x\rangle \sim K/2$, so in that limit the standard deviation scaled to the expectation value becomes $\sigma_x/\langle x\rangle \sim 1/\sqrt{3}\langle x\rangle$, which diverges as $\langle x\rangle \to 0$. For $K \gg 1$, we obtain $\sigma_x \sim K^{-1}$ and $\langle x\rangle \sim 1$, so $\sigma_x/\langle x\rangle \sim K^{-1} \to 0$ as $\langle x\rangle \to 1$. In that case the standard deviation vanishes: all dipoles are oriented along the field.

(2) **Thermodynamics of isotropic dispersions of hard rods.**

  (a) The free energy is given by $\beta F = N [\ln \rho\omega - 1 + \langle \ln P\rangle + \rho B]$, with $\omega = \omega(T, \mu_S)$ and μ_S the chemical potential of the solvent, as discussed in the preceding chapter. Hence, using the chain rule of differentiation, we obtain $\beta\Pi = -\partial\beta F/\partial V|_{N,T,\mu_S} = -N\left[((\rho\omega)^{-1} \times (-\rho\omega/V)) - \rho B/V\right] = \rho + B\rho^2$.

  (b) Along similar lines, we find $\beta\mu = \partial\beta F/\partial N|_{V,T,\mu_S} = \ln\rho\omega + \langle\ln P\rangle + \rho B$.

  (c) Using the results of (a) and (b), we find that $\partial\beta\mu/\partial V|_{T,N,\mu_S} = V^{-1} - 2\rho B/V = -\partial\beta\Pi/\partial V|_{T,N,\mu_S}$.

  (d) Using Eq. (4.17), and the equality $\langle\langle|\sin\gamma|\rangle\rangle' = \pi/4$ in the isotropic phase, we find for the compressibility factor $Z \equiv \beta\Pi/\rho$ in the isotropic phase $Z = \beta\Pi/\rho = 1 + B\rho = 1 + \pi L^2 D\rho/4 = 1 + \phi L/D$. So, the van 't Hoff law $Z = 1$ applies when $\phi L/D \ll 1$ or $\phi \ll D/L \ll 1$. It breaks down completely when $\phi L/D \gg 1$. The transition occurs near $\phi L/D = 4$ as discussed in the main text. Note that the relative contribution of the third virial scales as $C/B^2 \sim D/L$ so will not contribute significantly, even if the second virial correction is much larger than unity [3].

(3) **The spinodal of the isotropic phase of hard rods.**

  (a) If we insert $P(\vec{u}) = P$, with $P = 1/4\pi$ a properly normalised constant of $\vec{u}$, in Eq. (4.21), we obtain $-\ln 4\pi = \lambda - (8c/\pi) \times \pi/4$. Indeed, for an isotropic distribution, we have $\langle|\sin\gamma|\rangle' = \pi/4$ as we have seen in the previous exercise. We could have guessed this because the molecular field must be an invariant of the orientation in the isotropic phase. It follows that $\lambda = 2c - \ln 4\pi$.

  (b) The square of the norm $||P_2||$ of the second Legendre polynomial is equal to $||P_2||^2 \equiv \int_{-1}^{+1} dx P_2^2(x) = \int_{-1}^{+1} dx(3x^2 - 1)^2/4 = 2/5$ as followed by straightforward integration. For the order parameter, we obtain $\langle P_2(x)\rangle = \int_{-1}^{+1} dx[1/2 + \varepsilon P_2(x)]P_2(x) = \varepsilon||P_2||^2 = 2\varepsilon/5$ because $\int_{-1}^{+1} dx P_2(x) = 0$ as can be easily checked.

  (c) To first order in ε, we obtain by Taylor expansion $\ln(1/2 + \varepsilon P_2(x)) = \ln 1/2 + 2\varepsilon P_2(x) + \cdots$. Using the findings of (a) and (b), we conclude that $2\varepsilon P_2(x) = 2c \times ((5/8)P_2(x)) \times (2/5)\varepsilon$. This must be valid for all values of x. Hence, $c = 4$.

(d) For $c < 4$, $\Delta F > 0$ for all $S \neq 0$, implying $S = 0$ corresponds to the lowest free energy state. For $c > 4$, $\Delta F < 0$ for all $S \neq 0$, implying that $S = 0$ no longer corresponds to the lowest free energy state. In fact, $\partial^2 \Delta F / \partial S^2$ changes from positive to negative: the isotropic state changes from a free energy minimum to a (local) free energy maximum. The crossover at $c = 4$ must be the spinodal because for that concentration the condition for marginal stability, $\partial^2 \Delta F / \partial S^2 = 0$, holds.

5. Maier–Saupe Theory

(1) **Spinodal instability of the isotropic phase.**

(a) Define $f \equiv \beta \Delta F / N$, for which Eq. (5.5) becomes $f = \sigma - \beta \Delta \epsilon S^2 / 2$ with $\sigma \equiv \langle \ln 4\pi P(\vec{u}) \rangle$. Insert $P(\vec{u}) = P(x)/2\pi$ and $P(x) = 1/2 + \varepsilon P_2(x)$ in the expression for σ, Taylor expand the logarithm to quadratic order in ϵ, so make use of $\ln(1 + \delta) = \delta - \delta^2/2 + \cdots$ for $|\delta| \ll 1$, and recall that $\int_{-1}^{+1} dx \, P_2^2(x) = 2/5$, to obtain $\sigma \sim 2\varepsilon^2/5$. Using the equality $\varepsilon = 5S/2$ produces the given expression.

(b) The free energy has a minimum for $S = 0$ if $\beta \Delta \epsilon < 5$, and a maximum for that value if $\beta \Delta \epsilon > 5$, because $\partial^2 f / \partial S^2 = 5(1 - \beta \Delta \epsilon / 5)$ is positive for $\beta \Delta \epsilon < 5$ and negative for $\beta \Delta \epsilon > 5$.

(c) For the entropy, we have $\Delta S = -\partial \Delta F / \partial T = -5 N k_B S^2 / 2$. The energy becomes $\Delta U = \Delta F + T \Delta S = -N \Delta \epsilon S^2$. (Actually, this already follows by writing $\Delta F = \Delta U - T \Delta S$ and presuming $\Delta \epsilon$ does not depend on the temperature.) This implies that for $\beta \Delta \epsilon < 5$ the entropy loss must be greater than the gain in energy.

(d) For the integrals, it makes sense to make the co-ordinate transformation $y = (x - \langle x \rangle)/\sqrt{2}\sigma_x$. To calculate the variance, use the fact that $\int_{-\infty}^{+\infty} dy \, y^2 \exp(-ay^2) = -(d/da) \int_{-\infty}^{+\infty} dy \exp(-ay^2) = -(d/da)\sqrt{\pi/a} = \sqrt{\pi/a}/2a$. Alternatively, make use of $\int_{-\infty}^{+\infty} dy \, y^2 \exp(-y^2) = \int_{0}^{+\infty} dt \, t^{1/2} \times \exp(-t) = \Gamma(3/2) = \sqrt{\pi}/2$, with $\Gamma(\cdot)$ the Gamma function [4].

(e) According to the Boltzmann distribution, we have $P(S) \propto \exp -\beta \Delta F = \exp[-\frac{5}{2}N(1 - \frac{1}{5}\beta \Delta \epsilon)S^2]$. By comparing this with the Gaussian distribution, we conclude that the expectation value of the order parameter equals naught, $\langle S \rangle = 0$, and that for the variance we have Eq. (5.16). Notice that even though $S \in [-1/2, 1]$, the exponent scales as $N S^2$, and hence, the integration domain may be extended to plus and minus infinity in the thermodynamic limit $N \gg 1$. What we also find is that for fixed N, $\sigma_x^2 \to \infty$ if $\beta \Delta \epsilon \to 5$. Apparently, for temperatures near the spinodal temperature, order parameter fluctuations become very large indeed.

(2) **Solving the Maier–Saupe equation.**

(a) We have $P(\vec{u}) = P(x)/2\pi = Z^{-1}\exp(\beta\Delta\epsilon S P_2(x))$, so $S = \langle P_2(x)\rangle = Z^{-1}\int_{-1}^{+1} dx\, P_2(x)\exp(K P_2(x)) = Z^{-1}\int_{-1}^{+1} dx\,(\partial\exp(K P_2(x))/\partial K)$. Hence, $S = Z^{-1}(d/dK)\int_{-1}^{+1} dx\exp(K P_2(x)) = Z^{-1}dZ/dK = d\ln Z/dK$. Note that the integration of the azimuthal co-ordinate produces a constant that drops out of the equations.

(b) The integral can be done by the co-ordinate transformation $t \equiv x\sqrt{3K/2}$ and realising that $\exp(t^2)$ can be written as $\exp[-(it)^2]$ with $i = \sqrt{-1}$ the imaginary number.

(c) Define $L = L(K) \equiv 2^{-1}\int_{-1}^{+1} dx[3x^2 - 1]\exp 3Kx^2/\int_{-1}^{+1} dx\exp 3Kx^2$. Plot $S = L$ versus $\beta\Delta\epsilon = K/L$ for values of the dummy variable $K \in [-1.5, 5]$. The curve crosses the trivial solution $S = 0$ for $\beta\Delta\epsilon = 5$, so two solutions branch off at that point: one with positive order parameter and one with negative order parameter.

(d) In exercise (1), we defined $f \equiv \beta\Delta F/N$, for which Eq. (5.5) becomes $f = \sigma - \beta\Delta\epsilon S^2/2 = \sigma - KS/2$ with $\sigma \equiv \langle\ln(4\pi P(\vec{u}))\rangle = \langle\ln 2P(x)\rangle$. Inserting $P(x) = Z^{-1}\exp(K P_2(x))$, to give $f = \ln(4\pi/Z) + KS/2$ and $Z = \int_{-1}^{+1} dx\exp[K P_2(x)]$, now only integrates over the variable x.

(3) **The nematic–smectic A transition.**

(a) The first term describes the Gibbs entropy loss due to the positional ordering, and the second one is the excess molecular field a particle experiences from local ordering.

(b) Apply functional differentiation as explained in the main text, and introduce a Lagrange multiplier to enforce normalisation of the distribution function. So, replace $\psi(z)$ by $\psi(z) + \delta\psi(z)$ in the free energy Eq. (5.18), Taylor expand to first order in $\delta\psi(z)$, and write $\Delta F[\psi + \delta\psi] = F[\psi] + \int_{-d/2}^{+d/2} dz\delta\psi(z)[\delta F/\delta\psi]$. The term between the square brackets is the functional derivative. Fix the Lagrange multiplier to normalise the distribution.

(c) Insert the distribution $\psi = 1/d$, which is uniform and normalised. We have $\sigma = d^{-1}\int_{-d/2}^{+d/2} dz\cos(2\pi z/d) = (2\pi)^{-1}d^{-1}\int_{-\pi}^{+\pi} dy\cos y = 0$. The free energy, Eq. (5.18), is zero because $\ln(d^{-1}d) = 0$ and $\sigma = 0$.

(d) Insert the ansatz $\psi(z) = d^{-1}(1 + \varepsilon\cos(2\pi z/d))$ by way of a smectic perturbation to the uniform solution in the free energy Eq. (5.18), and perform a Taylor expansion to second order in the small parameter ε, analogous to what we did in exercise (1). This gives $\beta\Delta F/N = 2\sigma^2[1 - \beta\Delta\epsilon/4+\cdots]$ showing that for $\beta\Delta\epsilon > 4$ the smectic phase has a lower free energy than the nematic phase.

6. Beyond Maier–Saupe and Onsager

(1) **Lyotropic nematics of semi-flexible chains.**

(a) Balancing the orientational part of the free energy $L\langle P^{-1/2}\nabla_{\vec{u}}^2 P^{1/2}\rangle/2L_P$ $\propto L/L_P$ and the excluded volume part $\rho B \propto \rho L^2 D$ shows that $L/L_P \sim \rho L^2 D$ and that the nematic state must set in for $\rho L L_P D \sim 1$.

(b) Let $f = f(x)$ and $g = g(x)$ be two real-valued functions of $x \in [-1, +1]$, and then by successive partial integration, we find $\int_{-1}^{+1} dx f(x)\nabla_x^2 g(x) =$ $\int_{-1}^{+1} dx\, f(x)\frac{d}{dx}[(1-x^2)\frac{d}{dx}g(x)] = [f(1-x^2)\frac{d}{dx}]_{-1}^{+1} - \int_{-1}^{+1} dx(1-x^2)\frac{df}{dx}\frac{dg}{dx}$ $= -[g(1-x^2)\frac{df}{dx}]_{-1}^{+1} = +\int_{-1}^{+1} dx g \nabla_x^2 f$.

(c) How to work the variational derivative of the excluded volume term, we already encountered in Chap. 4, Eq. (4.11). Ditto for the contribution of the Lagrange parameter enforcing the normalisation of P, see Eq. (4.9). That of the conformational (orientational) term, Eq. (6.1), follows the same recipe by making appropriate use of a Taylor expansion. Define $f_{or}[P] \equiv \int d\vec{u} \sqrt{P(\vec{u})}\nabla_{\vec{u}}^2 \sqrt{P(\vec{u})}$. Then $f_{or}[P + \delta P] = \int d\vec{u} \sqrt{P + \delta P}\nabla_{\vec{u}}^2 \sqrt{P + \delta P}$ $= \int d\vec{u}[\sqrt{P} + \frac{\delta P}{2\sqrt{P}}]\nabla_{\vec{u}}^2[\sqrt{P} + \frac{\delta P}{2\sqrt{P}}] = f_{or}[P] + \int d\vec{u}[\frac{\delta P}{2\sqrt{P}}\nabla_{\vec{u}}^2\sqrt{P} +$ $\sqrt{P}\nabla_{\vec{u}}^2\frac{\delta P}{2\sqrt{P}}]$. Using the Hermitian property of the Laplacian operator $\nabla_{\vec{u}}^2$, see (b), we see that the second term in the square brackets becomes equal to the first. Hence, $f_{or}[P + \delta P] = f_{or}[P] + \int d\vec{u}\delta P[\frac{1}{\sqrt{P}}\nabla_{\vec{u}}^2 P]$ $\equiv f_{or}[P] + \int d\vec{u}\delta P[\frac{\delta f_{or}}{\delta P}]$. Hence, $\delta f_{or}/\delta P = \frac{1}{\sqrt{P}}\nabla_{\vec{u}}^2 P$. Collecting all terms gives Eq. (6.5).

(d) The term involving the Laplacian in Eq. (6.6) is zero for $P(x) = P$ a constant. The excluded volume term for constant P is $8c_p P$. Hence, $\lambda = 8c_p P$, noting that for a normalised distribution, $P = 1/2$.

(e) Note that according to Eq. (4.20), $\int_{-1}^{+1} dx' P(x') \int_0^{2\pi} d\phi|\sin\gamma|$ $=$ $\int_{-1}^{+1} dx' P(x')\, 2\pi\left[\frac{\pi}{4} - \frac{5\pi}{32}P_2(x)P_2(x')\right]$. Insert $P(x) = \frac{1}{2} + \varepsilon P_2(x)$, with $|\varepsilon| \ll 1$, to give $2\pi\left[\frac{\pi}{4} - \frac{5\pi}{32}P_2(x)\frac{2}{5}\varepsilon\right]$. Note further that in that case (to linear order in ε) $\frac{1}{\sqrt{P}}\nabla_x^2\sqrt{P} = -6\varepsilon P_2(x)$. Inserting this in Eq. (6.5) gives $-6\varepsilon P_2(x) = -\lambda + 24c_p - c_p\varepsilon P_2(x)$. Equating the constant terms and the terms proportional to $P_2(x)$ gives the desired result, using the fact that the result must be valid for all values of ε. Note that in a more formal way, we may make use of the orthogonality of the Legendre polynomials $P_0(x) = 1$ and $P_2(x)$.

(2) **Thermotropic nematics of semi-flexible chains.**

(a) Compare the molecular field for lyotropics of hard rod-like particles, $\beta U_{mol} = 2\rho\langle B(\gamma)\rangle'$, Eq. (4.13), with that for thermotropics, $\beta U_{mol} = \beta U_{iso} - \beta \Delta\varepsilon S P_2(\vec{n} \cdot \vec{u})$, Eq. (5.1). The constant term U_{iso} can be absorbed in λ. Realise further that $2\rho\langle B(\gamma)\rangle' = 2\rho L^2 D\langle|\sin\gamma|\rangle'$

becomes $2\rho L^2 D\langle|\sin\gamma|\rangle' \times 2L_P/L = 16c_p\langle|\sin\gamma|\rangle'/\pi$ for semi-flexible chains. Notice that here we used $P(\vec{u})$ in our definition of the angular brackets, rather than $P(x)$ as we did in exercise 1 (c). In the latter case, $\beta U_{mol} = 8c_p\langle|\sin\gamma|\rangle'/\pi^2$. Multiplying $\beta\Delta\epsilon S P_2(\vec{n}\cdot\vec{u})$ in a similar fashion by $2L_P/L$, we obtain from Eq. (6.5), Eq. (6.6).

(b) For isotropic solutions, $P = 1/2$ and $S = 0$, so from Eq. (6.6) we conclude that $\lambda = 0$ for the isotropic phase.

(c) Notice that for $P(x) = \frac{1}{2} + \varepsilon P_2(x)$, with ε a small positive or negative number, $P^{-1/2}\nabla_x^2 P^{1/2} = -6\varepsilon P_2(x)$. Further, in that case, $S = \int_{-1}^{+1} dx\, P_2(x)P(x) = 2\varepsilon/5$. Inserting this in Eq. (6.6), realising that the distribution is properly normalised, implying that $\lambda = 0$, gives the desired result, Eq. (6.7).

(d) The longer the chain, the stronger the interaction a chain has with neighbouring chains, presuming they are stretched along the nematic director due to the nematic ordering field.

(e) It implies that $T_{IN} \propto \Delta\epsilon L_P/L$. The larger L_P the stiffer the chain, the less bending fluctuations are suppressed, the more stable the nematic phase. Or, a stronger interaction of the chains in the nematic phase stabilises the nematic.

(3) **McMillan-type theory for hard-rod smectics.**

(a) If ϕ is the relevant concentration scale, then we expect to need to replace $\beta\Delta\epsilon$ by some constant times ϕ. The constant we set to retrieve the correct prediction of the nematic–smectic A transition for $\phi = 0.49$.

(b) No need to worry about chemical potentials and osmotic pressures because the transition within this prescription is continuous, so does not involve the co-existing phases.

References

1. D. Andrienko, *Introduction to liquid crystals*, J. Mol. Liq. **267** (2018), 520.
2. J.-P. Hansen and I.R. McDonald, *Theory of simple liquids - with applications to soft matter*, 4th edition (AP, Amsterdam, 2013).
3. T. Odijk, *Theory of lyotropic polymer liquid crystals*, Macromolecules **19** (1986), 2313.
4. M. Stone and P. Goldbart, *Mathematics for physics* (CUP, Cambridge, 2000).

Index

© The Author(s), under exclusive license to Springer Nature Switzerland AG 2022
P. van der Schoot, *Molecular Theory of Nematic (and Other) Liquid Crystals*,
SpringerBriefs in Physics, https://doi.org/10.1007/978-3-030-99862-2

Printed in the United States
by Baker & Taylor Publisher Services